INTRODUCTION TO CHEMICAL ANALYSIS

Introduction to CHEMICAL ANALYSIS

Barry Griffin

Oxford Book Company
Jaipur ▪ New Delhi

Published By

Oxford Book Company

Regd. Off. "Bony Residency", Gate No. 2
Opp. Tilak Public School, Vishweswariya Nagar,
Gopalpura Road, Jaipur - 302018, India
Ph: 0141 - 2761280, Telefax: 2761381
email - oxfordbookjpr@gmail.com
website - oxfordbookcompany.com

Delhi Office: 102, 1st Floor, Satyam House,
4327/3, Ansari Road, Darya Ganj,
New Delhi - 110002, India
Ph: 011-45652440

Distributed by
KEYA INTERNATIONAL PRESS LLP
E-mail: keyainternationalpressllp@gmail.com

First published in India, 2024

ISBN : 9789355240675

Typeset by Shivangi Computers, Jaipur

Printed and bound in India by Replika Press Pvt. Ltd.

Contents

Preface

In the fields of science, engineering, industry and statistics, the accuracy of a measurement system is the degree of closeness of measurements of a quantity to that quantity's actual value. The precision of a measurement system, also called reproducibility or repeatability, is the degree to which repeated measurements under unchanged conditions show the same results. Although The two words reproducibility and repeatability can be synonymous incolloquial use, they are deliberately contrasted in the context of the scientific method. A measurement system can be accurate but not precise, precise but not accurate, neither, or both. For example, if an experiment contains a systematic error, then increasing the sample size generally increases precision but does not improve accuracy. The end result would be a consistent yet inaccurate string of results from the flawed experiment. Eliminating the systematic error improves accuracy but does not change precision.

Ideally a measurement device is both accurate and precise, with measurements all close to and tightly clustered around the known value. The accuracy and precision of a measurement process is usually established by repeatedly measuring some traceable reference standard. Such standards are defined in the International System of Units and maintained by national standards organisations such as the National Institute of Standards and Technology. This also applies when measurements are repeated and averaged. In that case, the term standard error is properly applied: the precision of the average is equal to the known standard deviation of the process divided by the square root of the number of measurements averaged. Further, the central limit theorem shows that the probability distribution of the averaged measurements will be closer to a normal distribution than that of individual measurements.

The accuracy paradox for predictive analytics states that predictive models with a given level of accuracy may have greater predictive power than models with higher accuracy. It may be better to avoid the accuracy metric in favour of other metrics such as precision and recall. In situations where the minority class is more important, F-measure may be more appropriate, especially in situations with very skewed class imbalance. An alternate performance measure that treats both classes with equal importance is "balanced accuracy":

The book provide a clear understanding of the principles, applications and limitations of the various techniques involved in analytical chemistry.

— Barry Griffin

Chapter 1

X-ray Computed Tomography

X-ray computed tomography or Computer tomography (CT), is a medical imaging method employing tomography created by computer processing. Digital geometry processing is used to generate a three-dimensional image of the inside of an object from a large series of two-dimensional X-ray images taken around a single axis of rotation.

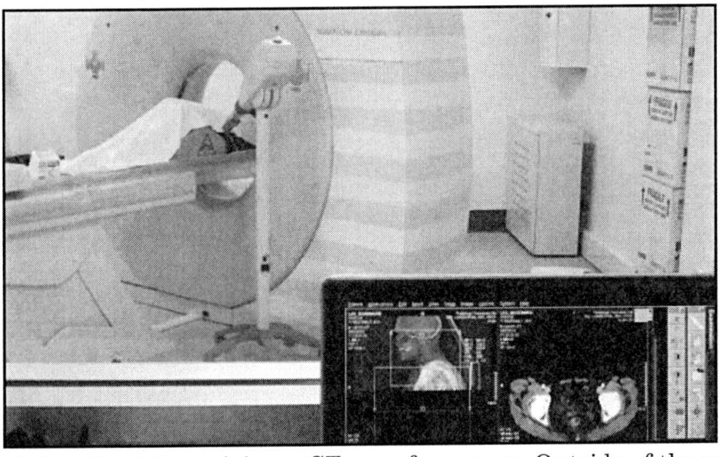

Figure 1: A patient is receiving a CT scan for cancer. Outside of the scanning room is an imaging computer that reveals a 3D image of the body's interior.

CT produces a volume of data that can be manipulated, through a process known as "windowing", in order to demonstrate various bodily structures based on their ability to block the X-ray beam. Although historically the images generated were in the axial or transverse plane, perpendicular to the long axis of the body, modern scanners allow this volume of data to be reformatted in various planes or even as volumetric (3D) representations of structures. Although most common in medicine, CT is also used in other fields, such as nondestructive materials testing. Another example is archaeological uses such as imaging the contents of sarcophagi.

Usage of CT has increased dramatically over the last two decades in many countries. An estimated 72 million scans were performed in the United States in 2007. It is estimated that 0.4% of current cancers in the United States are due to CTs performed in the past and that this may increase to as high as 1.5-2% with 2007 rates of CT usage; however, this estimate is disputed.

Entymology

The word "tomography" is derived from the Greek tomos (slice) and graphein (to write). Computed tomography was originally known as the "EMI scan" as it was developed at a research branch of EMI, a company best known today for its music and recording business. It was later known as computed axial tomography (CAT or CT scan) and body section röntgenography.

Although the term "computed tomography" could be used to describe positron emission tomography and single photon emission computed tomography, in practice it usually refers to the computation of tomography from X-ray images, especially in older medical literature and smaller medical facilities.

In MeSH, "computed axial tomography" was used from 1977–79, but the current indexing explicitly includes "X-ray" in the title.

Types of CT Machine

Spinning tube, commonly called spiral CT, in which an entire X-ray tube is spun around the central axis of the area being scanned. These are the dominant type of scanners on the market because they have been manufactured longer and offer lower cost of production and purchase. The main limitation of this type is the bulk and inertia of the equipment (X-ray tube assembly and detector array on the opposite side of the circle) which limits the speed at which the equipment can spin.

Electron beam tomography is a specific form of CT in which a large enough X-ray tube is constructed so that only the path of the electrons, traveling between the cathode and anode of the X-ray tube, are spun using deflection coils. This type has a major advantage since sweep speeds can be much faster, allowing for less blurry imaging of moving structures, such as the heart and arteries. However, far fewer CTs of this design have been produced, mainly due to the higher cost associated with building a much larger X-ray tube and detector array.

Diagnostic Use

Since its introduction in the 1970s, CT has become an important tool in medical imaging to supplement X-rays and medical ultrasonography. It has more recently been used for preventive medicine or screening for disease, for example CT colonography for patients with a high risk of colon cancer, or full-motion heart scans for patients with high risk of heart disease. A number of institutions offer full-body scans for the general population.

Head

Computed tomography of human brain, from base of the skull to top. Taken with intravenous contrast medium.

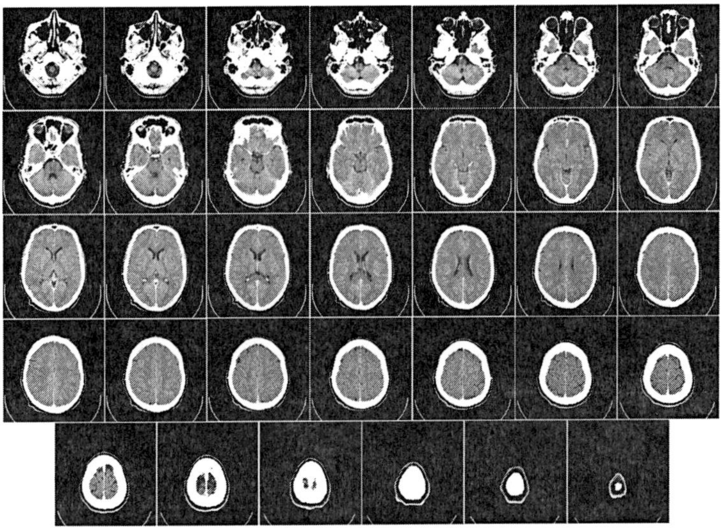

Figure 2: Computed tomography of human brain, from base of the skull to top. Taken with intravenous contrast medium.

Bone Reconstructed in 3D

CT scanning of the head is typically used to detect infarction, tumours, calcifications, haemorrhage and bone trauma. Of the above, hypodense (dark) structures indicate infarction or tumours, hyperdense (bright) structures indicate calcifications and haemorrhage and bone trauma can be seen as disjunction in bone windows. Ambulances equipped with small bore multi-sliced CT scanners respond to cases involving stroke or head trauma.

Lungs

CT can be used for detecting both acute and chronic changes in the lung parenchyma, that is, the internals of the lungs. It is particularly relevant here because normal two-dimensional X-rays do not show such defects. A variety of different techniques are used, depending on the suspected abnormality.

For evaluation of chronic interstitial processes (emphysema, fibrosis, and so forth), thin sections with high spatial frequency reconstructions are used; often scans are performed both in inspiration and expiration. This special technique is called high resolution CT. Therefore, it produces a sampling of the lung and not continuous images.

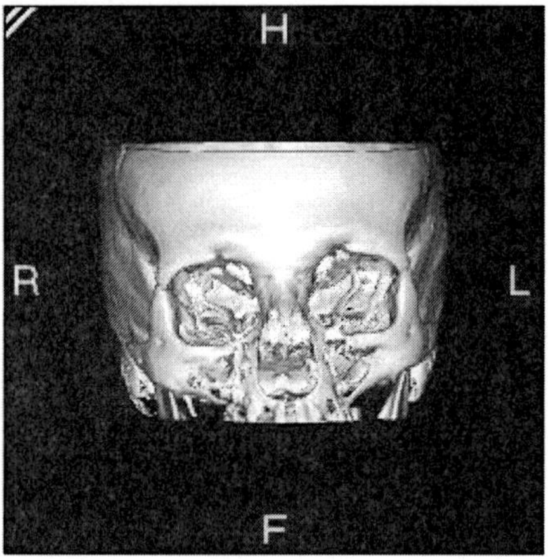

Figure 3: Bone reconstructed in 3D

Pulmonary Angiogram

Example of a CTPA, demonstrating a saddle embolus (dark horizontal line) occluding the pulmonary arteries (bright white triangle)

CT pulmonary angiogram (CTPA) is a medical diagnostic test used to diagnose pulmonary embolism (PE). It employs computed tomography to obtain an image of the pulmonary arteries.

Cardiac

With the advent of subsecond rotation combined with multi-slice CT (up to 320-slices), high resolution and high speed can be obtained at the same time, allowing excellent imaging of the coronary arteries (cardiac CT angiography).

Abdominal and Pelvic

CT Scan of 11 cm Wilms' tumour of right kidney in 13 month old patient. CT is a sensitive method for diagnosis of abdominal diseases. It is used frequently to determine stage of cancer and to follow progress. It is also a useful test to investigate acute abdominal pain.

Extremities

CT is often used to image complex fractures, especially ones around joints, because of its ability to reconstruct the area of interest in multiple planes. Fractures, ligamentous injuries and dislocations can easily be recognised with a 0.2 mm resolution.

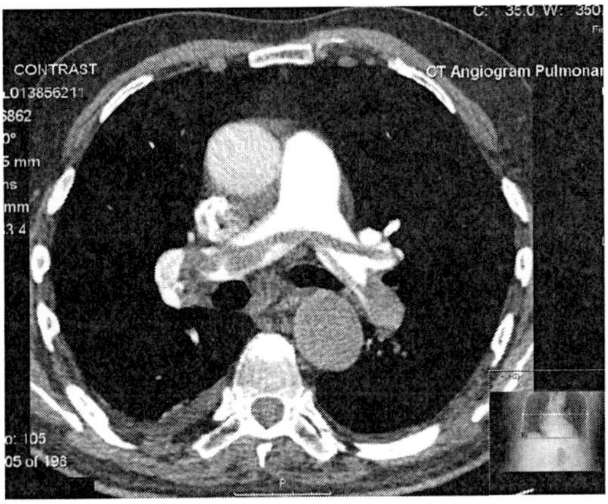

Figure 4: Example of a CTPA, demonstrating a saddle embolus (dark horizontal line) occluding the pulmonary arteries (bright white triangle)

Advantages

There are several advantages that CT has over traditional 2D medical radiography. First, CT completely eliminates the superimposition of images of structures outside the area of interest. Second, because of the inherent high-contrast resolution of CT, differences between tissues that differ in physical density by less than 1% can be distinguished. Finally, data from a single CT imaging procedure consisting of either multiple contiguous or one helical scan can be viewed as images in the axial, coronal, or sagittal planes, depending on the diagnostic task. This is referred to as multiplanar reformatted imaging.

CT is regarded as a moderate- to high-radiation diagnostic technique. The improved resolution of CT has permitted the development of new investigations, which may have advantages; compared to conventional radiography, for example, CT angiography avoids the invasive insertion of a catheter.

CT Colonography (also known as Virtual Colonoscopy or VC for short) may be as useful as a barium enema for detection of tumours, but may use a lower radiation dose. CT VC is increasingly being used in the UK as a diagnostic test for bowel cancer and can negate the need for a colonoscopy.

The radiation dose for a particular study depends on multiple factors: volume scanned, patient build, number and type of scan sequences, and desired resolution and image quality. In addition, two helical CT scanning parameters that can be adjusted easily and that have a profound effect on radiation dose are tube current and pitch. Computed tomography (CT) scan has been shown to be more accurate than radiographs in evaluating anterior interbody fusion but may still over-read the extent of fusion.

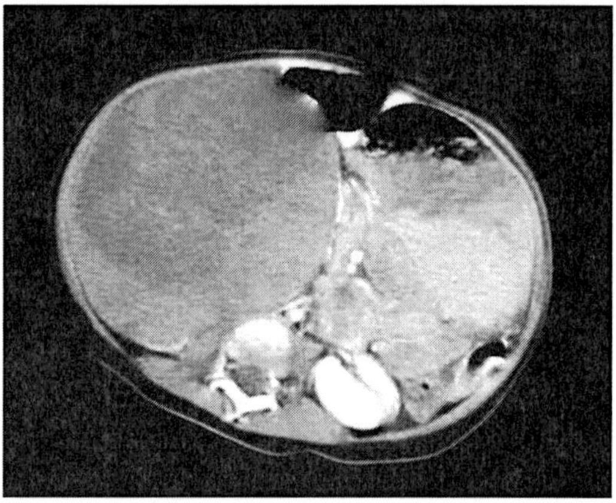

Figure 5: CT Scan of 11 cm Wilms' tumor of right kidney in 13 month old patient.

Adverse Effects

The increased use of CT scans has been the greatest in two fields: screening of adults (screening CT of the lung in smokers, virtual colonoscopy, CT cardiac screening, and whole-body CT in asymptomatic patients) and CT imaging of children. Shortening of the scanning time to around 1 second, eliminating the strict need for the subject to remain still or be sedated, is one of the main reasons for the large increase in the pediatric population (especially for the diagnosis of appendicitis). CT scans of children have been estimated to produce non-negligible increases in the probability of lifetime cancer mortality, leading to calls for the use of reduced current settings for CT scans of children.

These calculations are based on similar radiation exposures experienced by those present during the atomic bomb explosions in Japan during the second world war and nuclear industry works. Estimated lifetime cancer mortality risks attributable to the radiation exposure from a CT in a 1 year old are 0.18% (abdominal) and 0.07% (head) — an order of magnitude higher than for adults — although those figures still represent a small increase in cancer mortality over the background rate. In the United States, of approximately 600,000 abdominal and head CT examinations annually performed in children under the age of 15 years, a rough estimate is that 500 of these individuals might ultimately die from cancer attributable to the CT radiation.

The additional risk is still low: 0.35% compared to the background risk of dying from cancer of 23%. However, if these statistics are extrapolated to the current number of CT scans, the additional rise in cancer mortality could be 1.5 to 2%. Furthermore, certain conditions can require children to be exposed to multiple CT scans.

In 2009, a number of studies that further defined the risk of cancer that may be caused by CT scans appeared. One study indicated that radiation by CT scans is often higher and more variable than cited, and each of the 19,500 CT scans that are daily performed in the US is equivalent to 30 to 442 chest X-rays in radiation.

It has been estimated that CT radiation exposure will result in 29,000 new cancer cases just from the CT scans performed in 2007. The most common cancers caused by CT are thought to be lung cancer, colon cancer, and leukemia with younger people, and women more at risk. These conclusions, however, are criticized by the American College of Radiology (ACR), which maintains that the life expectancy of CT scanned patients is not that of the general population and that the model of calculating cancer is based on total-body radiation exposure and thus faulty.

CT scans can be performed with different settings for lower exposure in children, although these techniques are often not employed. Surveys have suggested that, at the current time, many CT scans are performed unnecessarily.

Ultrasound scanning or magnetic resonance imaging are alternatives (for example, in appendicitis or brain imaging) without the risk of radiation exposure.

Although CT scans come with an additional risk of cancer (it can be estimated that the radiation exposure from a full body scan is the same as standing 2.4 km away from the World War II atomic bomb blasts in Japan), especially in children, the benefits that stem from their use outweigh the risk in many cases.

Studies support informing parents of the risks of pediatric CT scanning.

Typical Scan Dose

Examination	*Typical effective dose (mSv)*	*(millirem)*
X-ray Personnel security screening scan	0.00025	0.025
Chest X-ray	0.1	10
Head CT	1.5	150
Screening mammography	3	300
Abdomen CT	5.3	530
Chest CT	5.8	580
CT colonography (virtual colonoscopy)	3.6–8.8	360–880
Chest, abdomen and pelvis CT	9.9	990
Cardiac CT angiogram	6.7-13	670–1300
Barium enema	15	1500
Neonatal abdominal CT	20	2000

Reactions to Contrast Agents

This section does not cite any references or sources. Please help improve this section by adding citations to reliable sources. Unsourced material may be challenged and removed. Because contrast CT scans rely on intravenously administered contrast agents in order to provide superior image quality, there is a low but non-negligible level of risk associated with the contrast agents themselves. Many patients report nausea and discomfort, including warmth in the crotch, which mimics the sensation of wetting oneself. Certain patients may experience severe and potentially life-threatening allergic reactions to the contrast dye.

The contrast agent may also induce kidney damage. The risk of this is increased with patients who have preexisting renal insufficiency, preexisting diabetes, or reduced intravascular volume. In general, if a patient has normal kidney function, then the risks of contrast nephropathy are negligible. Patients with mild kidney impairment are usually advised to ensure full hydration for several hours before and after the injection. For moderate kidney failure, the use of iodinated contrast should be avoided; this may mean using an alternative technique instead of CT, e.g., MRI. However, patients with severe renal failure requiring dialysis do not require special precautions, as their kidneys have so little function remaining that any further damage would not be noticeable and the dialysis will remove the contrast agent.

Low-dose CT Scan

An important issue within radiology today is how to reduce the radiation dose during CT examinations without compromising the image quality. In general, higher radiation doses result in higher-resolution images, while lower doses lead to increased image noise and unsharp images. Increased dosage raises the risk of radiation induced cancer — a four-phase abdominal CT gives the same radiation dose as 300 chest x-rays. Several methods that can reduce the exposure to ionizing radiation during a CT scan exist.

New software technology can significantly reduce the required radiation dose.

Individualise the examination and adjust the radiation dose to the body type and body organ examined. Different body types and organs require different amounts of radiation.

Prior to every CT examination, evaluate the appropriateness of the exam whether it is motivated or if another type of examination is more suitable. Higher resolution is not always suitable for any given scenario, such as detection of small pulmonary masses.

Artifacts

This section does not cite any references or sources. Please help improve this section by adding citations to reliable sources. Unsourced material may be challenged and removed. (September 2009)

Although CT is a relatively accurate test, it is liable to produce artifacts.

Streak Artifact

Streaks are often seen around materials that block most X-rays, such as metal or bone. These streaks can be caused by undersampling, photon starvation, motion, beam hardening, or scatter. This type of artifact commonly occurs in the posterior fossa of the brain, or if there are metal implants. The streaks can be reduced using newer reconstruction techniques.

Partial Volume Effect

This appears as "blurring" over sharp edges. It is due to the scanner being unable to differentiate between a small amount of high-density material (e.g., bone) and a larger amount of lower density (e.g., cartilage). The processor tries to average out the two densities or structures, and information is lost. This can be partially overcome by scanning using thinner slices.

Ring Artifact

Probably the most common mechanical artifact, the image of one or many "rings" appears within an image. This is usually due to a detector fault.

Noise Artifact

This appears as graining on the image and is caused by a low signal to noise ratio. This occurs more commonly when a thin slice thickness is used. It can also occur when the power supplied to the X-ray tube is insufficient to penetrate the anatomy.

Motion Artifact

This is seen as blurring and/or streaking, which is caused by movement of the object being imaged. Motion blurring might be reduced using a new technique called IFT (incompressible flow tomography).

Windmill

Streaking appearances can occur when the detectors intersect the reconstruction plane. This can be reduced with filters or a reduction in pitch.

Beam Hardening

This can give a "cupped appearance". It occurs when there is more attenuation in the center of the object than around the edge. This is easily corrected by filtration and software.

Prevalence

Usage of CT has increased dramatically over the last two decades. An estimated 72 million scans were performed in the United States in 2007. In Calgary, Canada 12.1% of people who present to the emergency with an urgent complaint received a CT scan, most commonly either of the head or of the abdomen. The percentage who received CT, however, varied markedly by the emergency physician who saw them from 1.8% to 25%. In the emergency department in the United States, CT or MRI imaging is done in 15% of people who present with injuries as of 2007 (up from 6% in 1998).

Process

3D reconstruction of the brain and eyeballs from CT scanned DICOM images. In this image, areas with the density of bone or air were made transparent, and the slices stacked up in an approximate free-space alignment. The outer ring of material around the brain are the soft tissues of skin and muscle on the outside of the skull. A black box encloses the slices to provide the black background. Since these are simply 2D images stacked up, when viewed on edge the slices disappear since they have effectively zero thickness. Each DICOM scan represents about 5mm of material averaged into a thin slice.

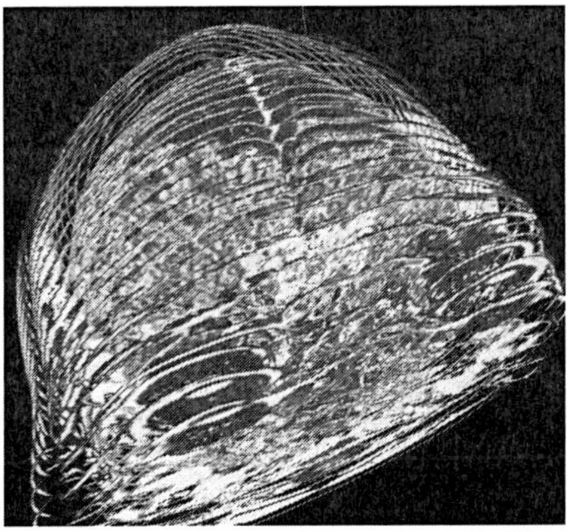

Figure 6: 3D reconstruction of the brain and eyeballs from CT scanned DICOM images. In this image, areas with the density of bone or air were made transparent, and the slices stacked up in an approximate free-space alignment. The outer ring of material around the brain are the soft tissues of skin and muscle on the outside of the skull. A black box encloses the slices to provide the black background. Since these are simply 2D images stacked up, when viewed on edge the slices disappear since they have effectively zero thickness. Each DICOM scan represents about 5mm of material averaged into a thin slice.

X-ray slice data is generated using an X-ray source that rotates around the object; X-ray sensors are positioned on the opposite side of the circle from the X-ray source. The earliest sensors were scintillation detectors, with photomultiplier tubes excited by (typically) cesium iodide crystals. Cesium iodide was replaced during the 1980s by ion chambers containing high-pressure Xenon gas. These systems were in turn replaced by scintillation systems based on photodiodes instead of photomultipliers and modern scintillation materials with more desirable characteristics. Many data scans are progressively taken as the object is gradually passed through the gantry.

Newer machines with faster computer systems and newer software strategies can process not only individual cross sections but continuously changing cross sections as the gantry, with the object to be imaged slowly and smoothly slid through the X-ray circle. These are called helical or spiral CT machines. Their computer systems integrate the data of the moving individual slices to generate three dimensional volumetric information (3D-CT scan), in turn viewable from multiple different perspectives on attached CT workstation monitors. This type of data acquisition requires enormous processing power, as the data are arriving in a continuous stream and must be processed in real-time.

In conventional CT machines, an X-ray tube and detector are physically rotated behind a circular shroud; in the electron beam tomography (EBT), the tube is far larger and higher power to support the high temporal resolution. The electron beam is deflected in a hollow funnel-shaped vacuum chamber. X-rays are generated when the beam hits the stationary target. The detector is also stationary. This arrangement can result in very fast scans, but is extremely expensive.

CT Scanner with Cover Removed to Show the Principle of Operation

CT is used in medicine as a diagnostic tool and as a guide for interventional procedures. Sometimes contrast materials such as intravenous iodinated contrast are used. This is useful to highlight structures such as blood vessels that otherwise would be difficult to delineate from their surroundings. Using contrast material can also help to obtain functional information about tissues. Once the scan data has been acquired, the data must be processed using a form of tomographic reconstruction, which produces a series of cross-sectional images. The most common technique in general use is filtered back projection, which is straightforward to implement and can be computed rapidly. In terms of mathematics, this method is based on the Radon transform. However, this is not the only technique available: the original EMI scanner solved the tomographic reconstruction problem by linear algebra, but this approach was limited by its high computational complexity, especially given the computer technology available at the time.

More recently, manufacturers have developed iterative physical model-based expectation-maximisation techniques. These techniques are advantageous because they use an internal model of the scanner's physical properties and of the physical laws of X-ray interactions. By contrast, earlier methods have assumed a perfect scanner and highly simplified physics, which leads to a number of artifacts and reduced resolution - the result is images with improved resolution, reduced noise and fewer artifacts, as well as the ability to greatly reduce the radiation dose in certain circumstances. The disadvantage is a very high computational requirement, which is at the limits of practicality for current scan protocols.

Pixels in an image obtained by CT scanning are displayed in terms of relative radiodensity. The pixel itself is displayed according to the mean attenuation of the tissue(s) that it corresponds to on a scale from +3071 (most attenuating) to -1024 (least attenuating) on the Hounsfield scale. Pixel is a two dimensional unit based on the matrix size and the field of view.

When the CT slice thickness is also factored in, the unit is known as a Voxel, which is a three-dimensional unit. The phenomenon that one part of the detector cannot differentiate between different tissues is called the "Partial Volume Effect".

That means that a big amount of cartilage and a thin layer of compact bone can cause the same attenuation in a voxel as hyperdense cartilage alone. Water has an attenuation of 0 Hounsfield units (HU), while air is -1000 HU, cancellous bone is typically +400 HU, cranial bone can reach 2000 HU or more (os temporale) and can cause artifacts.

The attenuation of metallic implants depends on atomic number of the element used: Titanium usually has an amount of +1000 HU, iron steel can completely extinguish the X-ray and is, therefore, responsible for well-known line-artifacts in computed tomograms. Artifacts are caused by abrupt transitions between low- and high-density materials, which results in data values that exceed the dynamic range of the processing electronics.

Contrast mediums used for X-ray CT, as well as for plain film X-ray, are called radiocontrasts. Radiocontrasts for X-ray CT are, in general, iodine-based.

Often, images are taken both with and without radiocontrast. CT images are called precontrast or native-phase images before any radiocontrast has been administrated, and postcontrast after radiocontrast administration.

Three-dimensional Reconstruction

Because contemporary CT scanners offer isotropic or near isotropic, resolution, display of images does not need to be restricted to the conventional axial images. Instead, it is possible for a software program to build a volume by "stacking" the individual slices one on top of the other. The program may then display the volume in an alternative manner.

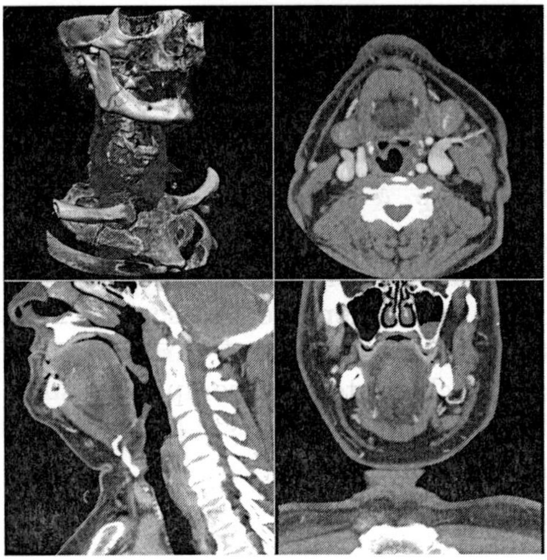

Figure 7: Typical screen layout for diagnostic software, showing one 3D and three MPR views

Multiplanar Reconstruction

Typical screen layout for diagnostic software, showing one 3D and three MPR views Multiplanar reconstruction (MPR) is the simplest method of reconstruction. A volume is built by stacking the axial slices. The software then cuts slices through the volume in a different plane (usually orthogonal). As an option, a special projection method, such as maximum-intensity projection (MIP) or minimum-intensity projection (MIP), can be used to build the reconstructed slices.

MPR is frequently used for examining the spine. Axial images through the spine will only show one vertebral body at a time and cannot reliably show the intervertebral discs. By reformatting the volume, it becomes much easier to visualise the position of one vertebral body in relation to the others.

Modern software allows reconstruction in non-orthogonal (oblique) planes so that the optimal plane can be chosen to display an anatomical structure. This may be particularly useful for visualising the structure of the bronchi as these do not lie orthogonal to the direction of the scan.

For vascular imaging, curved-plane reconstruction can be performed. This allows bends in a vessel to be "straightened" so that the entire length can be visualised on one image, or a short series of images. Once a vessel has been "straightened" in this way, quantitative measurements of length and cross sectional area can be made, so that surgery or interventional treatment can be planned.

MIP reconstructions enhance areas of high radiodensity, and so are useful for angiographic studies. MIP reconstructions tend to enhance air spaces so are useful for assessing lung structure.

3D Rendering Techniques

Surface rendering: A threshold value of radiodensity is set by the operator (e.g., a level that corresponds to bone). From this, a three-dimensional model can be constructed using edge detection image processing algorithms and displayed on screen. Multiple models can be constructed from various different thresholds, allowing different colours to represent each anatomical component such as bone, muscle, and cartilage. However, the interior structure of each element is not visible in this mode of operation.

Volume Rendering

Surface rendering is limited in that it will display only surfaces that meet a threshold density, and will display only the surface that is closest to the imaginary viewer. In volume rendering, transparency and colours are used to allow a better representation of the volume to be shown in a single image. For example, the bones of the pelvis could be displayed as semi-transparent, so that, even at an oblique angle, one part of the image does not conceal another.

Image Segmentation

Where different structures have similar radiodensity, it can become impossible to separate them simply by adjusting volume rendering parameters. The solution is called segmentation, a manual or automatic procedure that can remove the unwanted structures from the image.

Industrial Use

Industrial CT Scanning (industrial computed tomography) is a process which utilises x-ray equipment to produce 3D representations of components both externally and internally. Industrial CT scanning has been utilised in many areas of industry for internal inspection of components. Some of the key uses for CT scanning have been flaw detection, failure analysis, metrology, assembly analysis and reverse engineering applications

History

This section needs additional citations for verification. Please help improve this article by adding citations to reliable sources. Unsourced material may be challenged and removed. (September 2009) The prototype CT scanner A historic EMI-Scanner In the early

1900s, the Italian radiologist Alessandro Vallebona proposed a method to represent a single slice of the body on the radiographic film. This method was known as tomography. The idea is based on simple principles of projective geometry: moving synchronously and in opposite directions the X-ray tube and the film, which are connected together by a rod whose pivot point is the focus; the image created by the points on the focal plane appears sharper, while the images of the other points annihilate as noise. This is only marginally effective, as blurring occurs in only the "x" plane. There are also more complex devices that can move in more than one plane and perform more effective blurring.

In 1959, William Oldendorf, a UCLA neurologist and senior medical investigator at the West Los Angeles Veterans Administration hospital, conceived an idea for "scanning a head through a transmitted beam of X-rays, and being able to reconstruct the radiodensity patterns of a plane through the head" after watching an automated apparatus built to reject frost-bitten fruit by detecting dehydrated portions. In 1961, he built a prototype in which an X-ray source and a mechanically coupled detector rotated around the object to be imaged. By reconstructing the image, this instrument could get an X-ray picture of a nail surrounded by a circle of other nails, which made it impossible to X-ray from any single angle. In his landmark paper, published in 1961, he described the basic concept later used by Allan McLeod Cormack to develop the mathematics behind computerised tomography. In October, 1963 Oldendorf received a U.S. patent for a "radiant energy apparatus for investigating selected areas of interior objects obscured by dense material," Oldendorf shared the 1975 Lasker award with Hounsfield for that discovery .

Tomography has been one of the pillars of radiologic diagnostics until the late 1970s, when the availability of minicomputers and of the transverse axial scanning method – this last due to the work of Godfrey Hounsfield and South African-born Allan McLeod Cormack – gradually supplanted it as the modality of CT. In terms of mathematics, the method is based upon the use of the Radon Transform invented by Johann Radon in 1917. But as Cormack remembered later, he had to find the solution himself since it was only in 1972 that he learned of the work of Radon, by chance.

The first commercially viable CT scanner was invented by Sir Godfrey Hounsfield in Hayes, United Kingdom, at EMI Central Research Laboratories using X-rays. Hounsfield conceived his idea in 1967. The first EMI-Scanner was installed in Atkinson Morley Hospital in Wimbledon, England, and the first patient brain-scan was done on 1 October 1971. It was publicly announced in 1972.

The original 1971 prototype took 160 parallel readings through 180 angles, each 1° apart, with each scan taking a little over 5 minutes. The images from these scans took 2.5 hours to be processed by algebraic reconstruction techniques on a large computer. The scanner had a single photomultiplier detector, and operated on the Translate/ Rotate principle.

It has been claimed that thanks to the success of The Beatles, EMI could fund research and build early models for medical use. The first production X-ray CT machine (in fact called the "EMI-Scanner") was limited to making tomographic sections of the brain, but acquired the image data in about 4 minutes (scanning two adjacent slices), and the computation time (using a Data General Nova minicomputer) was about 7 minutes per picture. This scanner required the use of a water-filled Perspex tank with a pre-shaped rubber "head-cap" at the front, which enclosed the patient's head. The water-tank was used to reduce the dynamic range of the radiation reaching the detectors (between scanning outside the head compared with scanning through the bone of the skull). The images were relatively low resolution, being composed of a matrix of only 80 x 80 pixels.

In the U.S., the first installation was at the Mayo Clinic. As a tribute to the impact of this system on medical imaging the Mayo Clinic has an EMI scanner on display in the Radiology Department. Allan McLeod Cormack of Tufts University in Massachusetts independently invented a similar process, and both Hounsfield and Cormack shared the 1979 Nobel Prize in Medicine.

The first CT system that could make images of any part of the body and did not require the "water tank" was the ACTA (Automatic Computerised Transverse Axial) scanner designed by Robert S. Ledley, DDS, at Georgetown University. This machine had 30 photomultiplier tubes as detectors and completed a scan in only 9 translate/rotate cycles, much faster than the EMI-scanner. It used a DEC PDP11/34 minicomputer both to operate the servo-mechanisms and to acquire and process the images. The Pfizer drug company acquired the prototype from the university, along with rights to manufacture it. Pfizer then began making copies of the prototype, calling it the "200FS" (FS meaning Fast Scan), which were selling as fast as they could make them. This unit produced images in a 256×256 matrix, with much better definition than the EMI-Scanner's 80×80.

Since the first CT scanner, CT technology has vastly improved. Improvements in speed, slice count, and image quality have been the major focus primarily for cardiac imaging. Scanners now produce images much faster and with higher resolution enabling doctors to diagnose patients more accurately and perform medical procedures with greater precision. In the late 90's CT scanners have broke into two major groups, "Fixed CT" and "Portable CT". "Fixed CT Scanners" are large, require a dedicated power supply, electrical closet, HVAC cooling system, a separate workstation room, and a large lead lined room. "Fixed CT Scanners" can also be mounted inside large tractor trailers and driven from site to site and are known as "Mobile CT Scanners". "Portable CT Scanners" are light weight, small, and mounted on wheels. These scanners often have built-in lead shielding and run off of batteries or standard wall power.

Previous Studies

Pneumoencephalography Study for brain was quickly replaced by CT. A form of tomography can be performed by moving the X-ray source and detector during an exposure. Anatomy at the target level remains sharp, while structures at different levels are blurred. By varying the extent and path of motion, a variety of effects can be obtained, with variable depth of field and different degrees of blurring of "out of plane" structures:25 Although largely obsolete, conventional tomography is still used in specific situations such as dental imaging (orthopantomography) or in intravenous urography.

Tomosynthesis

Digital tomosynthesis combines digital image capture and processing with simple tube/detector motion as used in conventional radiographic tomography. Although there are some similarities to CT, it is a separate technique. In CT, the source/detector makes a complete 360-degree rotation about the subject obtaining a complete set of data from which images may be reconstructed. In digital tomosynthesis, only a small rotation angle (e.g. 40 degrees) with a small number of discrete exposures (e.g., 10) are used. This incomplete set of data can be digitally processed to yield images similar to conventional tomography with a limited depth of field. However, because the image processing is digital, a series of slices at different depths and with different thicknesses can be reconstructed from the same acquisition, saving both time and radiation exposure.

Because the data acquired are incomplete, tomosynthesis is unable to offer the extremely narrow slice widths that CT offers. However, higher resolution detectors can be used, allowing very high in-plane resolution, even if the Z-axis resolution is poor. The primary interest in tomosynthesis is in breast imaging, as an extension to mammography, where it may offer better detection rates with little extra increase in radiation exposure.

Reconstruction algorithms for tomosynthesis are significantly different from those of conventional CT because the conventional filtered back projection algorithm requires a complete set of data. Iterative algorithms based upon expectation maximisation are most commonly used, but are extremely computationally intensive. Some manufacturers have produced practical systems using off-the-shelf GPUs to perform the reconstruction.

Cyclic Voltammetry

Cyclic voltammetry or CV is a type of potentiodynamic electrochemical measurement. In a cyclic voltammetry experiment the working electrode potential is ramped linearly versus time like linear sweep voltammetry. Cyclic voltammetry takes the experiment a step further than linear sweep voltammetry which ends when it reaches a set potential. When cyclic voltammetry reaches a set potential, the working electrode's potential ramp is inverted. This inversion can happen multiple times during a single experiment. The current at the working electrode is plotted versus the applied voltage to give the cyclic voltammogram trace. Cyclic voltammetry is generally used to study the electrochemical properties of an analyte in solution.

Experimental Method

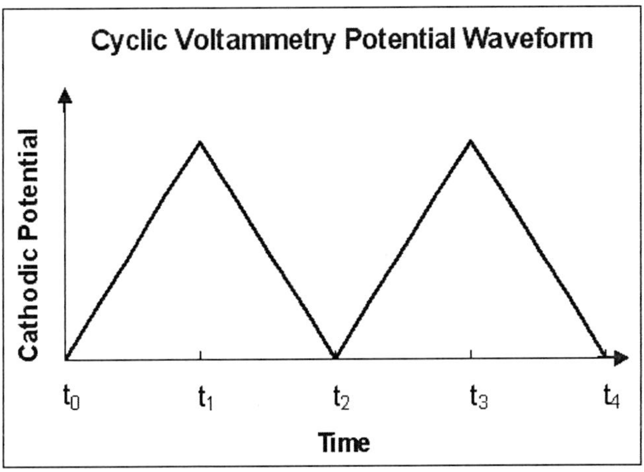

Figure 8: Cyclic voltammetry waveform

In cyclic voltammetry, the electrode potential ramps linearly versus time as shown. This ramping is known as the experiment's scan rate (V/s). The potential is applied between the reference electrode and the working electrode and the current is measured between the working electrode and the counter electrode. This data is then plotted as current (i) vs. potential (E). As the waveform shows, the forward scan produces a current peak for any analytes that can be reduced (or oxidised depending on the initial scan direction) through the range of the potential scanned. The current will increase as the potential reaches the reduction potential of the analyte, but then falls off as the concentration of the analyte is depleted close to the electrode surface. If the redox couple is reversible then when the applied potential is reversed, it will reach the potential that will reoxidise the product formed in the first reduction reaction, and produce a current of reverse polarity from the forward scan. This oxidation peak will usually have a similar shape to the reduction peak. As a result, information about the redox potential and electrochemical reaction rates of the compounds are obtained.

For instance if the electronic transfer at the surface is fast and the current is limited by the diffusion of species to the electrode surface, then the current peak will be proportional to the square root of the scan rate. This relationship is described by the Cottrell equation.

Characterisation

The utility of cyclic voltammetry is highly dependent on the analyte being studied. The analyte has to be redox active within the experimental potential window. It is also highly desirable for the analyte to display a reversible wave. A reversible wave is when an analyte is reduced or oxidised on a forward scan and is then reoxidised or rereduced in a predictable way on the return scan as shown in the first figure.

Even reversible couples contain polarisation overpotential and thus display a hysteresis between absolute potential between the reduction (E_{pc}) and oxidation peak (E_{pa}). This overpotential emerges from a combination of analyte diffusion rates and the intrinsic activation barrier of transferring electrons from an electrode to analyte. A theoretical description of polarisation overpotential is in part described by the Butler-Volmer equation and Cottrell equation. Conveniently in an ideal system the relationships reduces to, $|E_{pc} - E_{pa}| = \frac{57 \text{ mV}}{n}$, for an n electron process. Reversible couples will display a ratio of the peak currents passed at reduction (i_{pc}) and oxidation (i_{pa}) that is near unity ($1 = i_{pa}/i_{pc}$). This ratio can be perturbed for reversible couples in the presence of a following chemical reaction, stripping wave, or nucleation event.

When such reversible peaks are observed thermodynamic information in the form of half cell potential $E^0_{1/2}$ can be determined. When waves are semi-reversible such as when i_{pa}/i_{pc} is less than or greater than 1, it can be possible to determine even more information especially kinetic processes like following chemical reaction.

When waves are non-reversible it is impossible to determine what their thermodynamic $E^0_{1/2}$ is with cyclic voltammetry. This $E^0_{1/2}$ can be determined, however it often requires equal quantities of the analyte in both oxidation states. When a wave is non-reversible cyclic voltammetry can not determine if the wave is at its thermodynamic potential or shifted to a more extreme potential by some form of overpotential. The couple could be irreversible because of a following chemical process, a common example for transition metals is a shift in the geometry of the coordination sphere. If this is the case, then higher scan rates may show a reversible wave. It is also possible that the wave is irreversible due to a physical process most commonly some form of precipitation as discussed below. Some speculation can be made in regards to irreversible waves however they are generally outside the scope of cyclic voltammetry.

Experimental Setup

The method uses a reference electrode, working electrode, and counter electrode which in combination are sometimes referred to as a three-electrode setup. Electrolyte is usually added to the test solution to ensure sufficient conductivity. The combination of the solvent, electrolyte and specific working electrode material determines the range of the potential.

Electrodes are static and sit in unstirred solutions during cyclic voltammetry. This "still" solution method results in cyclic voltammetry's characteristic diffusion controlled peaks. This method also allows a portion of the analyte to remain after reduction or oxidation where it may display further redox activity. Stirring the solution between cyclic voltammetry traces is important as to supply the electrode surface with fresh analyte for each new experiment. The solubility of an analyte can change drastically with its overall charge. Since cyclic voltammetry usually alters the charge of the analyte it is common for reduced or oxidised analyte to precipitate out onto the electrode. This layering of analyte can insulate the electrode surface, display its own redox activity in subsequent scans, or at the very least alter the electrode surface. For this and other reasons it is often necessary to clean electrodes between scans.

Common materials for working electrodes include glassy carbon, platinum, and gold. These electrodes are generally encased in a rod of inert insulator with a disk exposed at one end. A regular working electrode has a radius within an order of magnitude of 1 mm. Having a controlled surface area with a defined shape is important for interpreting cyclic voltammetry results.

To run cyclic voltammetry experiments at high scan rates a regular working electrode is insufficient. High scan rates create peaks with large currents and increased resistances which result in distortions. Ultramicroelectrodes can be used to minimise the current and resistance.

The counter electrode, also known as the auxiliary or second electrode, can be any material which conducts easily and won't react with the bulk solution. Reactions occurring at the counter electrode surface are unimportant as long as it continues to conduct current well. To maintain the observed current the counter electrode will often oxidise or reduce the solvent or bulk electrolyte.

Variations

In some experiments an electroactive species is fixed to the surface of the electrode, for instance in microparticle voltammetry.

Potentiodynamic techniques also exist that add low-amplitude ac perturbation to a potential ramp and measure variable response in a single frequency (ac voltammetry) or in many frequencies simultaneously (potentiodynamic electrochemical impedance spectroscopy). The response in alternating current is two-dimensional – it is characterised by amplitude and phase. The amplitude and phase depend differently on frequency for constituents of ac response attributed to different processes (charge transfer, diffusion, double layer charging, etc.). Frequency response analysis enables simultaneous monitoring of the various processes that contribute to the potentiodynamic ac response of electrochemical system.

Distinctions

Cyclic voltammetry is not a hydrodynamic technique. In a hydrodynamic technique flow is achieved at the electrode surface by stirring the solution, pumping the solution, or rotating the electrode as is the case with rotating disk electrodes and rotating ring-disk electrodes. These techniques target steady state conditions which appear the same scanned from the positive or the negative, thus limiting them to linear sweep voltammetry.

Differential Scanning Calorimetry

Differential scanning calorimetry or DSC is a thermoanalytical technique in which the difference in the amount of heat required to increase the temperature of a sample and reference is measured as a function of temperature. Both the sample and reference are maintained at nearly the same temperature throughout the experiment. Generally, the temperature program for a DSC analysis is designed such that the sample holder temperature increases linearly as a function of time. The reference sample should have a well-defined heat capacity over the range of temperatures to be scanned.

The technique was developed by E.S. Watson and M.J. O'Neill in 1962, and introduced commercially at the 1963 Pittsburgh Conference on Analytical Chemistry and Applied Spectroscopy. The first adiabatic differential scanning calorimeter that could be used in biochemistry was developed by P.L. Privalov and D.R. Monaselidze in 1964. The term DSC was coined to describe this instrument which measures energy directly and allows precise measurements of heat capacity.

Detection of Phase Transitions

The basic principle underlying this technique is that when the sample undergoes a physical transformation such as phase transitions, more or less heat will need to flow to it than the reference to maintain both at the same temperature. Whether less or more heat must flow to the sample depends on whether the process is exothermic or endothermic. For example, as a solid sample melts to a liquid it will require more heat flowing to the sample to increase its temperature at the same rate as the reference. This is due to the absorption of heat by the sample as it undergoes the endothermic phase transition from solid to liquid. Likewise, as the sample undergoes exothermic processes (such as crystallisation) less heat is required to raise the sample temperature. By observing the difference in heat flow between the sample and reference, differential scanning calorimeters are able to measure the amount of heat absorbed or released during such transitions. DSC may also be used to observe more subtle phase changes, such as glass transitions. It is widely used in industrial settings as a quality control instrument due to its applicability in evaluating sample purity and for studying polymer curing.

DTA

An alternative technique, which shares much in common with DSC, is differential thermal analysis (DTA). In this technique it is the heat flow to the sample and reference that remains the same rather than the temperature. When the sample and reference are heated identically phase changes and other thermal processes cause a difference in temperature between the sample and reference. Both DSC and DTA provide similar information. Many modern commercial DTA are called heat flux DSC.

DSC Curves

The result of a DSC experiment is a curve of heat flux versus temperature or versus time. There are two different conventions: exothermic reactions in the sample shown with a positive or negative peak, depending on the kind of technology used in the experiment. This curve can be used to calculate enthalpies of transitions. This is done by integrating the peak corresponding to a given transition. It can be shown that the enthalpy of transition can be expressed using the following equation:

$$\Delta H = KA$$

where ΔH is the enthalpy of transition, K is the calorimetric constant, and A is the area under the curve. The calorimetric constant will vary from instrument to instrument, and can be determined by analysing a well-characterised sample with known enthalpies of transition.

Applications

Differential scanning calorimetry can be used to measure a number of characteristic properties of a sample. Using this technique it is possible to observe fusion and crystallisation events as well as glass transition temperatures T_g. DSC can also be used to study oxidation, as well as other chemical reactions. Glass transitions may occur as the temperature of an amorphous solid is increased. These transitions appear as a step in the baseline of the recorded DSC signal. This is due to the sample undergoing a change in heat capacity; no formal phase change occurs. As the temperature increases, an amorphous solid will become less viscous. At some point the molecules may obtain enough freedom of motion to spontaneously arrange themselves into a crystalline form. This is known as the crystallisation temperature (T_c).

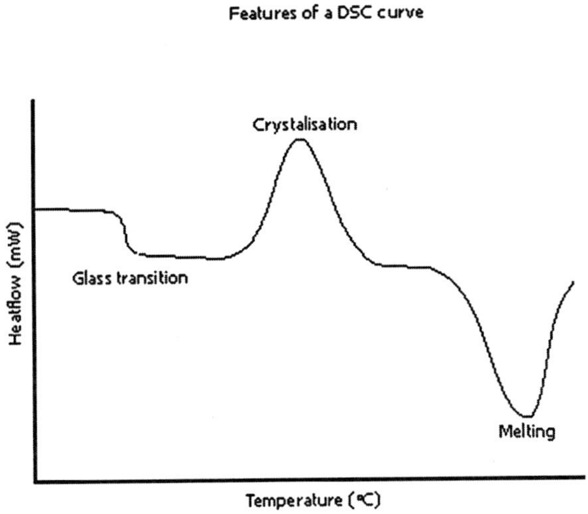

Figure 9: A schematic DSC curve demonstrating the appearance of several common features

This transition from amorphous solid to crystalline solid is an exothermic process, and results in a peak in the DSC signal. As the temperature increases the sample eventually reaches its melting temperature (T_m). The melting process results in an endothermic peak in the DSC curve. The ability to determine transition temperatures and enthalpies makes DSC a valuable tool in producing phase diagrams for various chemical systems.

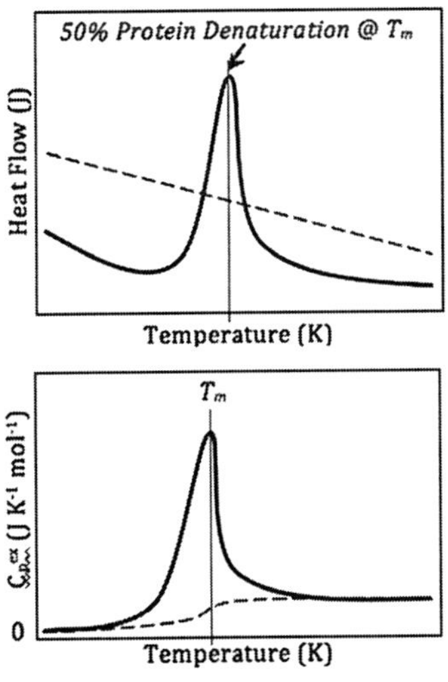

Figure 10: Top: A schematic DSC curve of amount of energy input (y) required to maintain each temperature (x), scanned across a range of temperatures. Bottom: Normalised curves setting the initial heat capacity as the reference. Buffer-buffer baseline (dashed) and protein-buffer variance (solid).

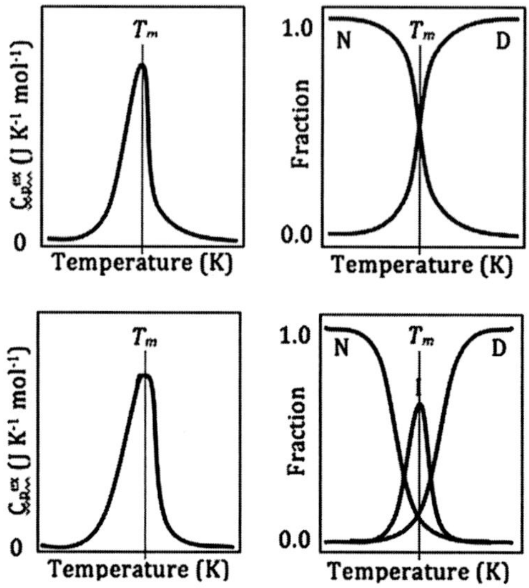

Figure 11: Normalised DSC curves using the baseline as the reference, and fractions of each conformational state (y) existing at each temperature, for two-state, and three-state proteins. Note the minuscule broadening in the peak of the three-state protein's DSC curve, which may or may not appear statistically significant to the naked eye.

Examples

The technique is widely used across a range of applications, both as a routine quality test and as a research tool. The equipment is easy to calibrate, using low melting indium at 156.5985 °C for example, and is a rapid and reliable method of thermal analysis.

Polymers

DSC is used widely for examining polymers to check their composition. Melting points and glass transition temperatures for most polymers are available from standard compilations, and the method can show possible polymer degradation by the lowering of the expected melting point, T_m, for example. T_m depends on the molecular weight of the polymer, so lower grades will have lower melting points than expected. The percentage crystallinity of a polymer can be found from the crystallisation peak of the DSC graph since the heat of fusion can be calculated from the area under an absorption peak. DSC can also be used to study thermal degradation of polymers. Impurities in polymers can be determined by examining thermograms for anomalous peaks, and plasticisers can be detected at their characteristic boiling points.

Liquid Crystals

DSC is used in the study of liquid crystals. As some forms of matter go from solid to liquid they go through a third state, which displays properties of both phases. This anisotropic liquid is known as a liquid crystalline or mesomorphous state. Using DSC, it is possible to observe the small energy changes that occur as matter transitions from a solid to a liquid crystal and from a liquid crystal to an isotropic liquid.

Oxidative Stability

Using differential scanning calorimetry to study the stability to oxidation of samples generally requires an airtight sample chamber. Usually, such tests are done isothermally (at constant temperature) by changing the atmosphere of the sample. First, the sample is brought to the desired test temperature under an inert atmosphere, usually nitrogen. Then, oxygen is added to the system. Any oxidation that occurs is observed as a deviation in the baseline. Such analysis can be used to determine the stability and optimum storage conditions for a material or compound.

Safety Screening

DSC makes a reasonable initial safety screening tool. In this mode the sample will be housed in a non-reactive crucible (often gold, or gold plated steel), and which will be able to withstand pressure (typically up to 100 bar). The presence of an exothermic event can then be used to assess the stability of a substance to heat. However, due to a combination of relatively poor sensitivity, slower than normal scan rates (typically 2-3 °/min - due to much heavier crucible) and unknown activation energy, it is necessary to deduct about 75-100 °C from the initial start of the observed exotherm to suggest a maximum temperature for the material. A much more accurate data set can be obtained from an adiabatic calorimeter, but such a test may take 2–3 days from ambient at a rate of a 3 °C increment per half hour.

Drug Analysis

DSC is widely used in the pharmaceutical and polymer industries. For the polymer chemist, DSC is a handy tool for studying curing processes, which allows the fine tuning of polymer properties. The cross-linking of polymer molecules that occurs in the curing process is exothermic, resulting in a positive peak in the DSC curve that usually appears soon after the glass transition.

In the pharmaceutical industry it is necessary to have well-characterised drug compounds in order to define processing parameters. For instance, if it is necessary to deliver a drug in the amorphous form, it is desirable to process the drug at temperatures below those at which crystallisation can occur.

General Chemical Analysis

Freezing-point depression can be used as a purity analysis tool when analysed by Differential scanning calorimetry. This is possible because the temperature range over which a mixture of compounds melts is dependent on their relative amounts. Consequently, less pure compounds will exhibit a broadened melting peak that begins at lower temperature than a pure compound.

Food Science

In food science research, DSC is used in conjunction with other thermal analytical techniques to determine water dynamics. Changes in water distribution may be correlated with changes in texture. Similar to materials science studies, the effects of curing on confectionery products can also be analysed.

Metals

In recent decades this technology has been involved in the study of metallic materials. Low melting materials such as solder are easily examined for example. It is known that it is possible to use DSC to find the solidus and liquidus temperature of a metal alloy, but at this time the widest application is the study of precipitations, and Guiner-Preston zones,

Chapter 2

Electron Paramagnetic Resonance

Electron paramagnetic resonance (EPR) or electron spin resonance (ESR) spectroscopy is a technique for studying chemical species that have one or more unpaired electrons, such as organic and inorganic free radicals or inorganic complexes possessing a transition metal ion. The basic physical concepts of EPR are analogous to those of nuclear magnetic resonance (NMR), but it is electron spins that are excited instead of spins of atomic nuclei. Because most stable molecules have all their electrons paired, the EPR technique is less widely used than NMR. However, this limitation to paramagnetic species also means that the EPR technique is one of great specificity, since ordinary chemical solvents and matrices do not give rise to EPR spectra.

EPR was first observed in Kazan State University by Soviet physicist Yevgeny Zavoisky in 1944, and was developed independently at the same time by Brebis Bleaney at the University of Oxford.

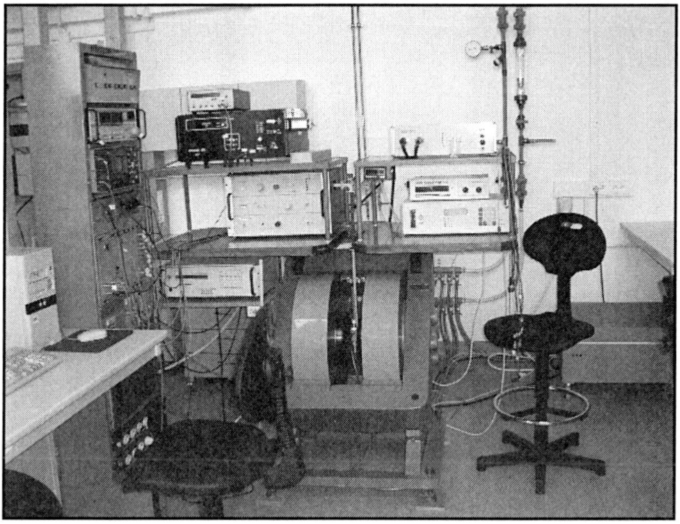

Figure 1: EPR spectrometer

Theory

Origin of an EPR Signal

Every electron has a magnetic moment and spin quantum number s = 1/2, with magnetic components $m_s = +1/2$ and $m_s = -1/2$. In the presence of an external magnetic field with strength B_0, the electron's magnetic moment aligns itself either parallel ($m_s = -1/2$) or antiparallel ($m_s = +1/2$) to the field, each alignment having a specific energy The parallel alignment corresponds to the lower energy state, and the separation between it and the upper state is $\Delta E = g_e\mu_B B_0$, where g_e is the electron's so-called g-factor and μ_B is the Bohr magneton. This equation implies that the splitting of the energy levels is directly proportional to the magnetic field's strength.

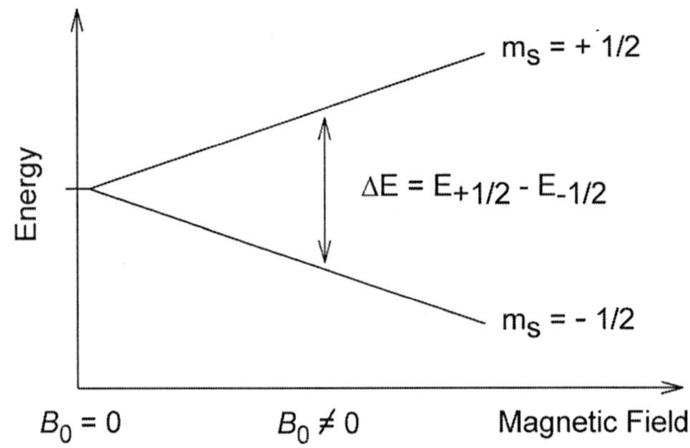

An unpaired electron can move between the two energy levels by either absorbing or emitting electromagnetic radiation of energy $\varepsilon = h\nu$ such that the resonance condition, = ΔE, is obeyed. Substituting in = $h\nu$ and $\Delta E = g_e\mu_B B_0$ leads to the fundamental equation of EPR spectroscopy: $h\nu = g_e\mu_B B_0$. Experimentally, this equation permits a large combination of frequency and magnetic field values, but the great majority of EPR measurements are made with microwaves in the 9000–10000 MHz (9–10 GHz) region, with fields corresponding to about 3500 G (0.35 T).

In principle, EPR spectra can be generated by either varying the photon frequency incident on a sample while holding the magnetic field constant or doing the reverse. In practice, it is usually the frequency that is kept fixed. A collection of paramagnetic centers, such as free radicals, is exposed to microwaves at a fixed frequency.

By increasing an external magnetic field, the gap between the $m_s = +1/2$ and $m_s = -1/2$ energy states is widened until it matches the energy of the microwaves, as represented by the double-arrow in the diagram above. At this point the unpaired electrons can move between their two spin states. Since there typically are more electrons in the lower state, due to the Maxwell-Boltzmann distribution, there is a net absorption of energy, and it is this absorption that is monitored and converted into a spectrum.

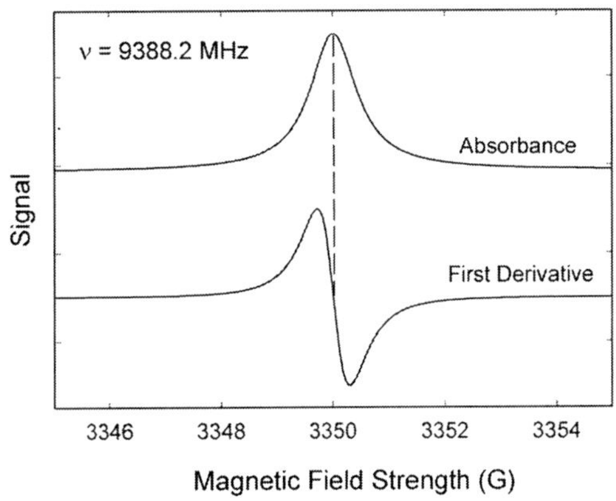

As an example of how $h\nu = g_e\mu_B B_0$ can be used, consider the case of a free electron, which has $g_e = 2.0023$, and the simulated spectrum shown at the right in two different forms. For the microwave frequency of 9388.2 MHz, the predicted resonance position is a magnetic field of about $B_0 = h\nu / g_e\mu_B = 0.3350$ tesla = 3350 gauss, as shown. Note that while two forms of the same spectrum are presented in the figure, most EPR spectra are recorded and published only as first derivatives.

Because of electron-nuclear mass differences, the magnetic moment of an electron is substantially larger than the corresponding quantity for any nucleus, so that a much higher electromagnetic frequency is needed to bring about a spin resonance with an electron than with a nucleus, at identical magnetic field strengths. For example, for the field of 3350 G shown at the right, spin resonance occurs near 9388.2 MHz for an electron compared to only about 14.3 MHz for ^{1}H nuclei. (For NMR spectroscopy, the corresponding resonance equation is $h\nu = g_N\mu_N B_0$ where g_N and μ_N depend on the nucleus under study.)

Maxwell-Boltzmann Distribution

In practice, EPR samples consist of collections of many paramagnetic species, and not single isolated paramagnetic centers.

If the population of radicals is in thermodynamic equilibrium, its statistical distribution is described by the Maxwell-Boltzmann equation

$$\frac{n_{\text{upper}}}{n_{\text{lower}}} = \exp\left(-\frac{E_{\text{upper}} - E_{\text{lower}}}{kT}\right) = \exp\left(-\frac{\Delta E}{kT}\right) = \exp$$

$$\left(-\frac{ò}{kT}\right) = \exp\left(-\frac{h\nu}{kT}\right)$$

where n_{upper} is the number of paramagnetic centers occupying the upper energy state, k is the Boltzmann constant, and T is the temperature in kelvins. At 298 K, X-band microwave frequencies (ν H– 9.75 GHz) give $n_{\text{upper}} / n_{\text{lower}}$ H" 0.998, meaning that the upper energy level has a smaller population than the lower one. Therefore, transitions from the lower to the higher level are more probable than the reverse, which is why there is a net absorption of energy.

The sensitivity of the EPR method (i.e., the minimum number of detectable spins N_{min}) depends on the photon frequency ν according to

$$N_{\text{min}} = \frac{k_1 V}{Q_0 k_f \nu^2 P^{1/2}}$$

where k_1 is a constant, V is the sample's volume, Q_0 is the unloaded quality factor of the microwave cavity (sample chamber), k_f is the cavity filling coefficient, and P is the microwave power in the spectrometer cavity. With k_f and P being constants, $N_{\text{min}} \sim (Q_0\nu^2)^{-1}$, i.e., $N_{\text{min}} \sim \nu^{-\text{a}}$, where **a** H– 1.5. In practice, **a** can change varying from 0.5 to 4.5 depending on spectrometer characteristics, resonance conditions, and sample size. In other words, the higher the spectrometer frequency the lower the detection limit (N_{min}), meaning greater sensitivity.

Spectral Parameters

In real systems, electrons are normally not solitary, but are associated with one or more atoms. There are several important consequences of this:

1. An unpaired electron can gain or lose angular momentum, which can change the value of its g-factor, causing it to differ from g_e. This is especially significant for chemical systems with transition-metal ions.
2. If an atom with which an unpaired electron is associated has a non-zero nuclear spin, then its magnetic moment will affect the electron. This leads to the phenomenon of hyperfine

coupling, analogous to J-coupling in NMR, splitting the EPR resonance signal into doublets, triplets and so forth.

3. Interactions of an unpaired electron with its environment influence the shape of an EPR spectral line. Line shapes can yield information about, for example, rates of chemical reactions.
4. The g-factor and hyperfine coupling in an atom or molecule may not be the same for all orientations of an unpaired electron in an external magnetic field. This anisotropy depends upon the electronic structure of the atom or molecule (e.g., free radical) in question, and so can provide information about the atomic or molecular orbital containing the unpaired electron.

The g Factor

Knowledge of the g-factor can give information about a paramagnetic center's electronic structure. An unpaired electron responds not only to a spectrometer's applied magnetic field B_0 but also to any local magnetic fields of atoms or molecules. The effective field B_{eff} experienced by an electron is thus written

$$B_{eff} = B_0(1 - \sigma)$$

where σ includes the effects of local fields (σ can be positive or negative). Therefore, the $h\nu = g_e\mu_B B_{eff}$ resonance condition (above) is rewritten as follows:

$$h\nu = g_e\mu_B B_{eff} = g_e\mu_B B_0(1 - \sigma)$$

The quantity $g_e(1 - \sigma)$ is denoted g and called simply the g-factor, so that the final resonance equation becomes

$$h\nu = g\mu_B B_0$$

This last equation is used to determine g in an EPR experiment by measuring the field and the frequency at which resonance occurs. If g does not equal g_e the implication is that the ratio of the unpaired electron's spin magnetic moment to its angular momentum differs from the free electron value. Since an electron's spin magnetic moment is constant (approximately the Bohr magneton), then the electron must have gained or lost angular momentum through spin-orbit coupling. Because the mechanisms of spin-orbit coupling are well understood, the magnitude of the change gives information about the nature of the atomic or molecular orbital containing the unpaired electron.

In general, the g factor is not a number but a second-rank tensor represented by nine numbers arranged in a 3×3 matrix. The principal axes of this tensor are determined by the local fields, for example, by the local atomic arrangement around the unpaired spin in a solid or in a molecule. Choosing an appropriate coordinate system (say, x,y,z) allows to "diagonalize" this tensor thereby reducing the maximum number of its components from nine to three, g_{xx}, g_{yy} and g_{zz}. For a single spin experiencing only Zeeman interaction with an external magnetic field, the position of the EPR resonance is given by the expression $g_{xx}B_x + g_{yy}B_y + g_{zz}B_z$. Here B_x, B_y and B_z are the components of the magnetic field vector in the coordinate system (x,y,z); their magnitudes change as the field is rotated, so as the frequency of the resonance. For a large ensemble of randomly oriented spins, the EPR spectrum consists three peaks of characteristic shape at frequencies $g_{xx}B_0$, $g_{yy}B_0$ and $g_{zz}B_0$: the low-frequency peak is positive in first-derivative spectra, the high-frequency peak is negative, and the central peak is bipolar. Such situation is commonly observed in powders and the spectra are therefore called "powder-pattern spectra". In crystals, the number of EPR lines is determined by the number of crystallographically equivalent orientations of the EPR spin (called "EPR center").

Hyperfine Coupling

Since the source of an EPR spectrum is a change in an electron's spin state, it might be thought that all EPR spectra for a single electron spin would consist of one line. However, the interaction of an unpaired electron, by way of its magnetic moment, with nearby nuclear spins, results in additional allowed energy states and, in turn, multi-lined spectra. In such cases, the spacing between the EPR spectral lines indicates the degree of interaction between the unpaired electron and the perturbing nuclei. The hyperfine coupling constant of a nucleus is directly related to the spectral line spacing and, in the simplest cases, is essentially the spacing itself.

Two common mechanisms by which electrons and nuclei interact are the Fermi contact interaction and by dipolar interaction. The former applies largely to the case of isotropic interactions (independent of sample orientation in a magnetic field) and the latter to the case of anisotropic interactions (spectra dependent on sample orientation in a magnetic field). Spin polarisation is a third mechanism for interactions between an unpaired electron and a nuclear spin, being especially important for π-electron organic radicals, such as the benzene radical anion. The symbols "*a*" or "*A*" are used for isotropic hyperfine coupling constants while "*B*" is usually employed for anisotropic hyperfine coupling constants.

In many cases, the isotropic hyperfine splitting pattern for a radical freely tumbling in a solution (isotropic system) can be predicted.

- For a radical having M equivalent nuclei, each with a spin of I, the number of EPR lines expected is $2MI + 1$. As an example, the methyl radical, CH_3, has three ^1H nuclei each with $I = 1/2$, and so the number of lines expected is $2MI + 1 = 2(3)(1/2) + 1 = 4$, which is as observed.
- For a radical having M_1 equivalent nuclei, each with a spin of I_1, and a group of M_2 equivalent nuclei, each with a spin of I_2, the number of lines expected is $(2M_1I_1 + 1)(2M_2I_2 + 1)$. As an example, the methoxymethyl radical, $H_2C(OCH_3)$, has two equivalent ^1H nuclei each with $I = 1/2$ and three equivalent ^1H nuclei each with $I = 1/2$, and so the number of lines expected is $(2M_1I_1 + 1)(2M_2I_2 + 1) = [2(2)(1/2) + 1][2(3)(1/2) + 1] = [3][4] = 12$, again as observed.

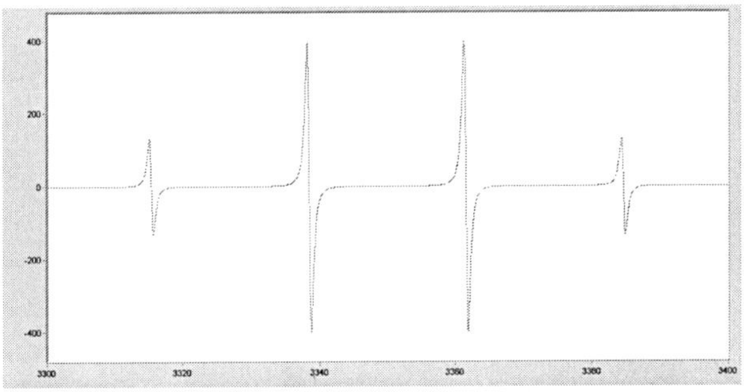

Figure 2: Simulated EPR spectrum of the CH_3 radical

- The above can be extended to predict the number of lines for any number of nuclei.

While it is easy to predict the number of lines a radical's EPR spectrum should show, the reverse problem, unravelling a complex multi-line EPR spectrum and assigning the various spacings to specific nuclei, is more difficult.

In the oft-encountered case of $I = 1/2$ nuclei (e.g., ^1H, ^{19}F, ^{31}P), the line intensities produced by a population of radicals, each possessing M equivalent nuclei, will follow Pascal's triangle. For example, the spectrum at the right shows that the three ^1H nuclei of the CH_3 radical give rise to $2MI + 1 = 2(3)(1/2) + 1 = 4$ lines with a 1:3:3:1 ratio. The line spacing gives a hyperfine coupling constant of $a_H = 23\ G$ for each of the three ^1H nuclei. Note again that the lines in this spectrum are *first derivatives* of absorptions.

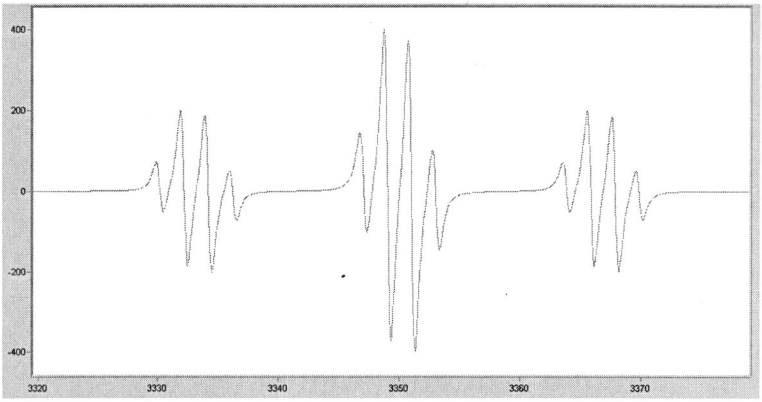

Figure 3: Simulated EPR spectrum of the $H_2C(OCH_3)$ radical

As a second example, consider the methoxymethyl radical, $H_2C(OCH_3)$. The two equivalent methyl hydrogens will give an overall 1:2:1 EPR pattern, each component of which is further split by the three methoxy hydrogens into a 1:3:3:1 pattern to give a total of 3×4 = 12 lines, a triplet of quartets. A simulation of the observed EPR spectrum is shown at the right, and agrees with the 12-line prediction and the expected line intensities.

Note that the smaller coupling constant (smaller line spacing) is due to the three methoxy hydrogens, while the larger coupling constant (line spacing) is from the two hydrogens bonded directly to the carbon atom bearing the unpaired electron. It is often the case that coupling constants decrease in size with distance from a radical's unpaired electron, but there are some notable exceptions, such as the ethyl radical (CH_2CH_3).

Resonance Linewidth Definition

Resonance linewidths are defined in terms of the magnetic induction B, and its corresponding units, and are measured along the x axis of an EPR spectrum, from a line's center to a chosen reference point of the line. These defined widths are called halfwidths and possess some advantages: for asymmetric lines values of left and right halfwidth can be given. The halfwidth ΔB_h is the distance measured from the line's center to the point in which absorption value has half of maximal absorption value in the center of resonance line. First inclination width $\Delta B_{1/2}$ is a distance from center of the line to the point of maximal absorption curve inclination. In practice, a full definition of linewidth is used. For symmetric lines, halfwidth$\Delta B_{1/2}$ = $2\Delta B_h$, and full inclination width $\Delta B_{max} = 2\Delta B_{1s}$

Pulsed EPR

The dynamics of electron spins are best studied with pulsed measurements . Microwave pulses typically 10–100 ns long are used to control the spins in the Bloch sphere. The spin-lattice relaxation time can be measured with an inversion recovery experiment.

As with pulsed NMR, the Hahn echo is central to many pulsed EPR experiments. A Hahn echo decay experiment can be used to measure the dephasing time, as shown in the animation below. The size of the echo is recorded for different spacings of the two pulses. This reveals the decoherence, which is not refocused by the π pulse. In simple cases, an exponential decay is measured, which is described by the T_2 time.

Applications

EPR spectroscopy is used in various branches of science, such as chemistry and physics, for the detection and identification of free radicals and paramagnetic centers such as F centers. EPR is a sensitive, specific method for studying both radicals formed in chemical reactions and the reactions themselves. For example, when frozen water (solid H_2O) is decomposed by exposure to high-energy radiation, radicals such as H, OH, and HO_2 are produced. Such radicals can be identified and studied by EPR. Organic and inorganic radicals can be detected in electrochemical systems and in materials exposed to UV light. In many cases, the reactions to make the radicals and the subsequent reactions of the radicals are of interest, while in other cases EPR is used to provide information on a radical's geometry and the orbital of the unpaired electron.

Medical and biological applications of EPR also exist. Although radicals are very reactive, and so do not normally occur in high concentrations in biology, special reagents have been developed to spin-label molecules of interest. These reagents are particularly useful in biological systems. Specially-designed nonreactive radical molecules can attach to specific sites in a biological cell, and EPR spectra can then give information on the environment of these so-called spin-label or spin-probes.

A type of dosimetry system has been designed for reference standards and routine use in medicine, based on EPR signals of radicals from irradiated polycrystalline **a**-alanine(the alanine deamination radical, the hydrogen abstraction radical, and the $(CO^{-}(OH))=C(CH_3)NH_2^{+}$ radical). This method is suitable for measuring gamma and x-rays, electrons, protons, and high-linear energy transfer (LET) radiation of doses in the 1 Gy to 100 kGy range.

EPR spectroscopy can be applied only to systems in which the balance between radical decay and radical formation keeps the free-radicals concentration above the detection limit of the spectrometer used. This can be a particularly severe problem in studying reactions in liquids. An alternative approach is to slow down reactions by studying samples held at cryogenic temperatures, such as 77 K (liquid nitrogen) or 4.2 K (liquid helium). An example of this work is the study of radical reactions in single crystals of amino acids exposed to x-rays, work that sometimes leads to activation energies and rate constants for radical reactions.

The study of radiation-induced free radicals in biological substances (for cancer research) poses the additional problem that tissue contains water, and water (due to its electric dipole moment) has a strong absorption band in the microwaveregion used in EPR spectrometers.

EPR also has been used by archaeologists for the dating of teeth. Radiation damage over long periods of time creates free radicals in tooth enamel, which can then be examined by EPR and, after proper calibration, dated. Alternatively, material extracted from the teeth of people during dental procedures can be used to quantify their cumulative exposure to ionizing radiation. People exposed to radiation from the Chernobyl disaster have been examined by this method.

Radiation-sterilised foods have been examined with EPR spectroscopy, the aim being to develop methods to determine if a particular food sample has been irradiated and to what dose.

Because of its high sensitivity, EPR was used recently to measure the quantity of energy used locally during a mechanochemical milling process.

EPR spectroscopy has been used to measure properties of crude oil, in particular asphaltene and vanadium content. EPR measurement of asphaltene content is a function of spin density and solvent polarity. Prior work dating to the 1960s has demonstrated the ability to measure vanadium content to sub-ppm levels.

High-field High-frequency Measurements

High-field high-frequency EPR measurements are sometimes needed to detect subtle spectroscopic details. However, for many years the use of electromagnets to produce the needed fields above 1.5 T was impossible, due principally to limitations of traditional magnet materials. The first multifunctional millimeter EPR spectrometer with a superconducting solenoid was described in the early 1970s by Prof. Y. S. Lebedev's group (Russian Institute of Chemical Physics,Moscow) in collaboration with L. G. Oranski's group (Ukrainian Physics and Technics Institute, Donetsk), which began working in the Institute of Problems of Chemical Physics, Chernogolovka around 1975. Two decades later, a W-band EPR spectrometer was produced as a small commercial line by the German Bruker Company, initiating the expansion of W-band EPR techniques into medium-sized academic laboratories. Today there still are only a few scientific centers in the world capable of high-field high-frequency EPR; among them are the Grenoble High Magnetic Field Laboratory in Grenoble, France, the Physics Department in Freie Universitαt Berlin, the National High Magnetic Field Laboratory in Tallahassee, US, the National Center for Advanced ESR Technology (ACERT) at Cornell University in Ithaca, US, the Department of Physiology, and Biophysics at Albert Einstein College of Medicine, Bronx, NY, the IFW in Dresden,Germany, the Institute of Physics of Complex Matter in Lausanne in Switzerland, and the Institute of Physics of the Leiden University, Netherlands.

The EPR waveband is stipulated by the frequency or wavelength of a spectrometer's microwave source.

EPR experiments often are conducted at X and, less commonly, Q bands, mainly due to the ready availability of the necessary microwave components (which originally were developed for radar applications). A second reason for widespread X and Q band measurements is that electromagnets can reliably generate fields up to about 1 tesla. However, the low spectral resolution over g-factor at these wavebands limits the study of paramagnetic centers with comparatively low anisotropic magnetic parameters. Measurements at $\nu > 40$ GHz, in the millimeter wavelength region, offer the following advantages:

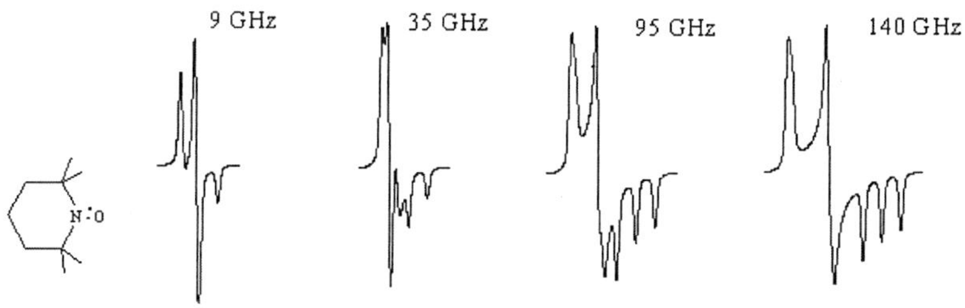

Figure 4: EPR spectra of a nitroxide radical as a function of frequency. Note the improvement in resolution from left to right.

1. EPR spectra are simplified due to the reduction of second-order effects at high fields.
2. Increase in orientation selectivity and sensitivity in the investigation of disordered systems.
3. The informativity and precision of pulse methods, e.g., ENDOR also increase at high magnetic fields.
4. Accessibility of spin systems with larger zero-field splitting due to the larger microwave quantum energy $h\nu$.
5. The higher spectral resolution over g-factor, which increases with irradiation frequency ν and external magnetic field B_0. This is used to investigate the structure, polarity, and dynamics of radical microenvironments in spin-modified organic and biological systems through the spin label and probe method.
6. Saturation of paramagnetic centers occurs at a comparatively low microwave polarising field B_1, due to the exponential dependence of the number of excited spins on the radiation frequency ν. This effect can be successfully used to study the relaxation and dynamics of paramagnetic centers as well as of superslow motion in the systems under study.
7. The cross-relaxation of paramagnetic centers decreases dramatically at high magnetic fields, making it easier to obtain more-precise and more-complete information about the system under study.

Fourier Transform Spectroscopy

Fourier transform spectroscopy is a measurement technique whereby spectra are collected based on measurements of the coherence of a radiative source, using time-domain or space-domain measurements of the electromagnetic radiation or other type of radiation. It can be applied to a variety of types of spectroscopy including optical spectroscopy, infrared spectroscopy (FTIR, FT-NIRS), nuclear magnetic resonance (NMR) and magnetic resonance spectroscopic imaging (MRSI), mass spectrometry and electron spin resonance spectroscopy. There are several methods for measuring the temporal coherence of the light, including the continuous wave *Michelson* or *Fourier transform* spectrometer and the pulsed Fourier transform spectrograph (which is more sensitive and has a much shorter sampling time than conventional spectroscopic techniques, but is only applicable in a laboratory environment).

The term *Fourier transform spectroscopy* reflects the fact that in all these techniques, a Fourier transform is required to turn the raw data into the actual spectrum, and in many of the cases in optics involving interferometers, is based on theWiener–Khinchin theorem.

Measuring an Emission Spectrum

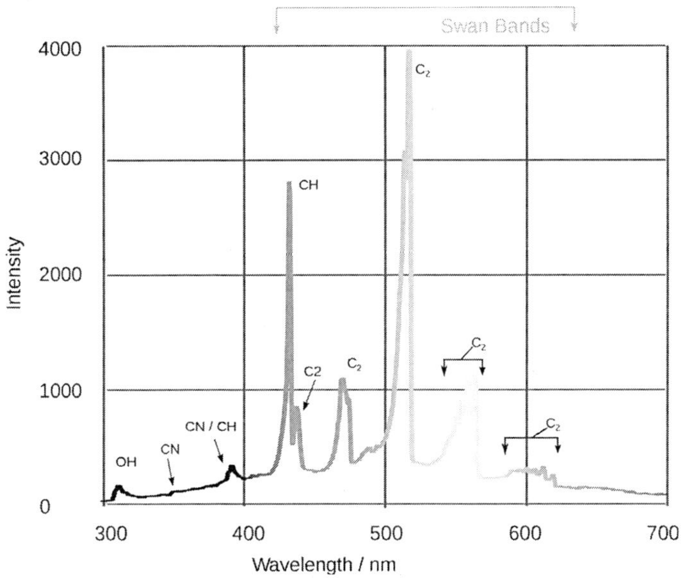

An example of a spectrum: The spectrum of light emitted by the blue flame of a butane torch. The horizontal axis is the wavelength of light, and the vertical axis represents how much light is emitted by the torch at that wavelength.

One of the most basic tasks in spectroscopy is to characterise the spectrum of a light source: How much light is emitted at each different wavelength. The most straightforward way to measure a spectrum is to pass the light through a monochromator, an instrument that blocks all of the light *except* the light at a certain wavelength (the unblocked wavelength is set by a knob on the monochromator). Then the intensity of this remaining (single-wavelength) light is measured. The measured intensity directly indicates how much light is emitted at that wavelength. By varying the monochromator's wavelength setting, the full spectrum can be measured. This simple scheme in fact describes how *some* spectrometers work.

Fourier transform spectroscopy is a less intuitive way to get the same information. Rather than allowing only one wavelength at a time to pass through to the detector, this technique lets through a beam containing many different wavelengths of light at once, and measures the *total* beam intensity. Next, the beam is modified to contain a *different* combination of wavelengths, giving a second data point. This process is repeated many times. Afterwards, a computer takes all this data and works backwards to infer how much light there is at each wavelength.

To be more specific, between the light source and the detector, there is a certain configuration of mirrors that allows some wavelengths to pass through but blocks others (due to wave interference). The beam is modified for each new data point by moving one of the mirrors; this changes the set of wavelengths that can pass through. As mentioned, computer processing is required to turn the raw data (light intensity for each mirror position) into the desired result (light intensity for each wavelength). The processing required turns out to be a common algorithm called the Fourier transform (hence the name, "Fourier transform spectroscopy"). The raw data is sometimes called an "interferogram". Because of the existing computer equipment requirements, and the ability of light to analyse very small amounts of substance, it is often beneficial to automate many aspects of the sample preparation. The sample can be better preserved and the results are much easier to replicate. Both of these benefits are important in testing situations that may later involve legal action, such as those involving drug specimens.

Measuring an Absorption Spectrum

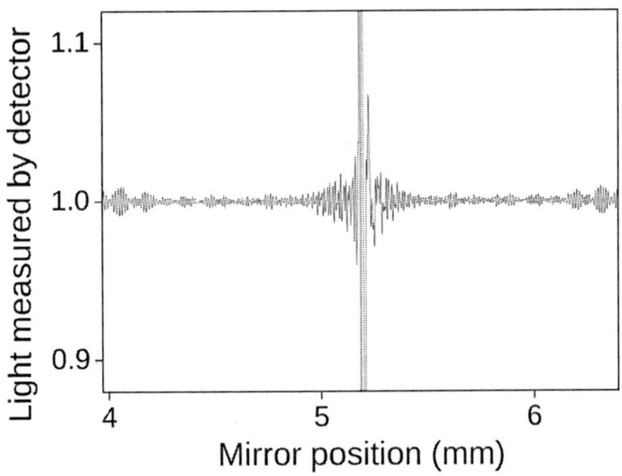

Figure 5: An "interferogram" from a Fourier transform spectrometer. This is the "raw data" which can be Fourier transformed into an actual spectrum. The peak at the center is the ZPD position ("Zero Path Difference"): Here, all the light passes through the interferometer because its two arms have equal length.

The method of Fourier transform spectroscopy can also be used for absorption spectroscopy. The primary example is "FTIR Spectroscopy", a common technique in chemistry.

In general, the goal of absorption spectroscopy is to measure how well a sample absorbs or transmits light at each different wavelength. Although absorption spectroscopy and emission spectroscopy are different in principle, they are closely related in practice; any technique for emission spectroscopy can also be used for absorption spectroscopy. First, the emission spectrum of abroadband lamp is measured (this is called the "background spectrum"). Second, the emission spectrum of the same lamp *shining through the sample* is measured (this is called the "sample spectrum"). The sample will absorb some of the light, causing the spectra to be different. The ratio of the "sample spectrum" to the "background spectrum" is directly related to the sample's absorption spectrum.

Accordingly, the technique of "Fourier transform spectroscopy" can be used both for measuring emission spectra (for example, the emission spectrum of a star), *and* absorption spectra (for example, the absorption spectrum of a glass of liquid).

Continuous Wave Michelson or Fourier Transform Spectrograph

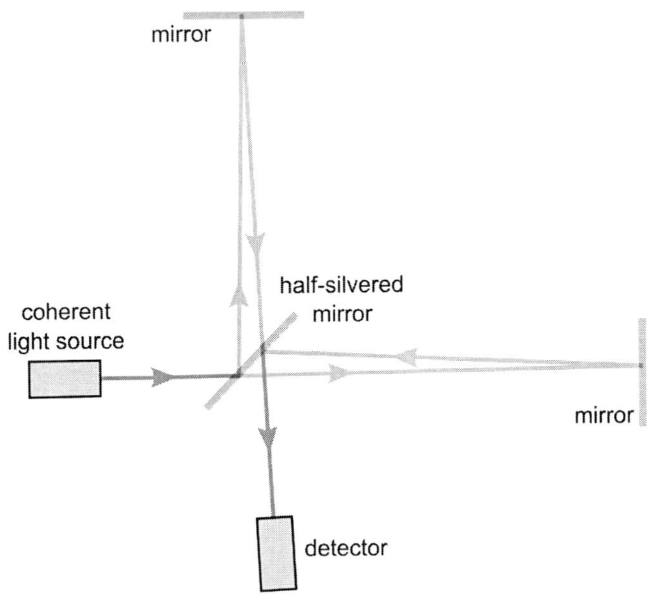

Figure 6: The Fourier transform spectrometer is just a Michelson interferometer but one of the two fully reflecting mirrors is movable, allowing a variable delay (in the travel-time of the light) to be included in one of the beams.

The Michelson spectrograph is similar to the instrument used in the Michelson-Morley experiment. Light from the source is split into two beams by a half-silvered mirror, one is reflected off a fixed mirror and one off a moving mirror which introduces a time delay—the Fourier transform spectrometer is just a Michelson interferometer with a movable mirror. The beams interfere, allowing the temporal coherence of the light to be measured at each different time delay setting, effectively converting the time domain into a spatial coordinate. By making measurements of the signal at many discrete positions of the moving mirror, the spectrum can be reconstructed using a Fourier transform of the temporal coherence of the light. Michelson spectrographs are capable of very high spectral resolution observations of very bright sources. The Michelson or Fourier transform spectrograph was popular for infrared applications at a time when infrared astronomy only had single pixel detectors. Imaging Michelson spectrometers are a possibility, but in general have been supplanted by imaging Fabry–Pérot instruments which are easier to construct.

Extracting the Spectrum

The intensity as a function of the path length difference (also denoted as retardation) in the interferometer p and wavenumber

$\tilde{\nu} = 1/\lambda$ is

$$I(p,\tilde{\nu}) = I(\tilde{\nu})[1+\cos(2\pi\tilde{\nu}p)],$$

where $I(\tilde{\nu})$ is the spectrum to be determined. Note that it is not necessary for $I(\tilde{\nu})$ to be modulated by the sample before the interferometer. In fact, most FTIR spectrometers place the sample after the interferometer in the optical path. The total intensity at the detector is

$$I(p) = \int_0^\infty I(p,\tilde{\nu})d\tilde{\nu} = \int_0^\infty I(\tilde{\nu})[1+\cos(2\pi\tilde{\nu}p)]d\tilde{\nu}.$$

This is just a Fourier cosine transform. The inverse gives us our desired result in terms of the measured quantity $I(p)$:

$$I(\tilde{\upsilon}) = 4\int_0^\omega [I(P) - \frac{1}{2}l(P=0)]COS(2\pi\tilde{\upsilon}p)dp.$$

Pulsed* Fourier Transform *Spectrometer

A pulsed Fourier transform *spectrometer does not employ transmittance techniques. In the most general description of pulsed FT spectrometry, a sample is exposed to an energizing event which causes a periodic response. The frequency of the periodic response, as governed by the field conditions in the spectrometer, is indicative of the measured properties of the analyte.*

Examples of Pulsed Fourier Transform Spectrometry

In magnetic spectroscopy (EPR, NMR), an RF pulse in a strong ambient magnetic field is used as the energizing event. This turns the magnetic particles at an angle to the ambient field, resulting in gyration. The gyrating spins then induce a periodic current in a detector coil. Each spin exhibits a characteristic frequency of gyration (relative to the field strength) which reveals information about the analyte.

In Fourier transform mass spectrometry, the energizing event is the injection of the charged sample into the strong electromagnetic field of a cyclotron. These particles travel in circles, inducing a current in a fixed coil on one point in their circle. Each travelling particle exhibits a characteristic cyclotron frequency-field ratio revealing the masses in the sample.

Free Induction Decay

Pulsed FT spectrometry gives the advantage of requiring a single, time-dependent measurement which can easily deconvolute a set of similar but distinct signals. The resulting composite signal, is called a *free induction decay,* because typically the signal will decay due to inhomogeneities in sample frequency, or simply unrecoverable loss of signal due to entropic loss of the property being measured.

Stationary forms of Fourier Transform Spectrometers

In addition to the scanning forms of Fourier transform spectrometers, there are a number of stationary or self-scanned forms. While the analysis of the interferometric output is similar to that of the typical scanning interferometer, significant differences apply, as shown in the published analyses. Some stationary forms retain the Fellgett multiplex advantage, and their use in the spectral region where detector noise limits apply is similar to the scanning forms of the FTS. In the photon-noise limited region, the application of stationary interferometers is dictated by specific consideration for the spectral region and the application.

Fellgett Advantage

One of the most important advantages of Fourier transform spectroscopy was shown by P.B. Fellgett, an early advocate of the method. The Fellgett advantage, also known as the multiplex principle, states that when obtaining a spectrum when measurement noise is dominated by detector noise (which is independent of the power of radiation incident on the detector), a multiplex spectrometer such as a Fourier transform spectrometer will produce a relative improvement in signal-to-noise ratio, compared to an equivalent scanning monochromator, of the order of the square root of m, where m is the number of sample points comprising the spectrum. However if the detector is shot noise dominated, the noise will be proportional to the square root of the power, thus for a broad boxcar spectrum (continuous broadband source), the noise is proportional to the square root of m, thus precisely offset the Fellgett's advantage. Shot noise is the main reason Fourier Transform Spectrometry was never popular for UV and visible spectrum.

Converting Spectra from Time Domain to Frequency Domain

$$S(t) = \int_{-\infty}^{\infty} I(\nu)e^{-i\nu 2\pi t}\, d\nu$$

The sum is performed over all contributing frequencies to give a signal S(t) in the time domain.

$$I(\nu) = \int_{-\infty}^{\infty} S(t)e^{i\nu 2\pi t}\, dt$$

gives non-zero value when S(t) contains a component that matches the oscillating function.

Remember that

$$e^{ix} = \cos x + i \sin x$$

Chapter 3

Forensic Chemistry

Forensic chemistry is the application of chemistry to law enforcement or the failure of products or processes. Many different analytical methods may be used to reveal what chemical changes occurred during an incident, and so help reconstruct the sequence of events. "Forensic chemistry is unique among chemical sciences in that its research, practice, and presentation must meet the needs of both the scientific and the legal communities. As such, forensic chemistry research is applied and derivative by nature and design, and it emphasizes metrology and validation."

Methods

One useful method is the gas chromatograph-mass spectrometer (GCMS), which is actually two instruments that are attached. The gas chromatograph is essentially a very hot oven holding a hollow coiled column. A drug sample is diluted in a solvent (e.g., chloroform, methanol) and is injected into this column, the solvent will evaporate very quickly leaving the drug to travel through the column. Different substances are retained in the column for different amounts of time.

The retention time, as compared to a known standard sample using the same method(same column length/polarity, same flow rate, same temperature program), can help to provide a positive identification for the presence of a compound of interest. The column eluent is then fed into a mass spectrometer.

A mass spectrometer bombards the eluant with electrons, causing it to fragment into ions. These ions are separated by their mass, commonly with the use of a quadrupole mass analyser or quadrupole ion trap, and detected by an electron multiplier. This provides a fragmentation pattern, which functions as a sort of fingerprint for each compound, and is compared to a reference sample.

Spectroscopy

Another instrument used to aide in identification of compounds is the Fourier Transform infrared spectrophotometer (FTIR). The sample is bombarded with infrared radiation. Polar bonds found in organic compounds have a natural frequency of vibration similar to the frequency of infrared radiation. When the frequency of the infrared radiation matches the natural frequency of the bond, the amplitude of the vibration increases, and the infrared is absorbed. The output of an infrared spectrophotometer charts the amount of light absorbed vs. the wavelength, typically with units of percent transmission andwavenumbers(cm^{-1}). Because both the frequency and the intensity of absorption are dependent on the type of bond, a skilled chemist can determine the functional groups present by examining the infrared spectrum.

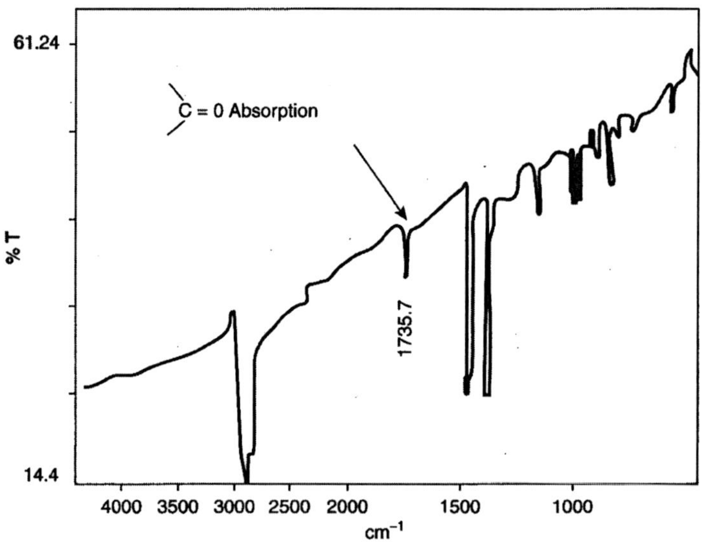

Figure 1: IR spectrum showing carbonylabsorption due to oxidative degradation of polypropylene

As with the GCMS the FTIR spectrum can be compared to that of a known sample, thus providing evidence for the identification of a compound. Spectroscopy can also help to identify materials used in failed products, especially polymers, additives and fillers. Samples can be taken by dissolution, or by cutting a thin slice using a microtome from the specimen under examination. Surfaces can be examined using Attenuated total reflectance spectroscopy, and the method has also been adapted to the optical microscope with infrared microspectroscopy. Ultraviolet-visible-near infrared spectroscopy is used to test for certain drugs of abuse. UV-visible-NIR microspectrophotometers are instruments able to measure the spectra of microscopic samples.

The UV-visible-NIR microspectrophotometer is used to compare known and questioned samples of trace evidence such as fibers and paint chips. They are also used in the analysis of inks and papers of questioned documents and to measure the colour of microscopic glass fragments. As these samples are not altered, UV-visible-NIR microspectroscopy is considered a non-destructive technique.

Thermoplastics can be analysed using characterisation techniques such as infrared spectroscopy, ultraviolet–visible spectroscopy, nuclear magnetic resonance spectroscopy, and an environmental scanning electron microscope. Failed samples can either be dissolved in a suitable solvent and examined directly (UV, IR and NMR spectroscopy) or be a thin film cast from solvent or cut using microtomy from the solid product.

Infrared spectroscopy is especially useful for assessing oxidation of polymers, such as the polymer degradation caused by faulty injection moulding. The spectrum shows the characteristic carbonyl group produced by oxidation of polypropylene, which made the product brittle. It was a critical part of a crutch, and when it failed, the user fell and injured herself very seriously. The spectrum was obtained from a thin film cast from a solution of a sample of the plastic taken from the failed forearm crutch.

Sample Integrity

Forensic chemists usually perform their analytical work in a sterile laboratory decreasing the risk of sample contamination. In order to prevent tampering, forensic chemists must keep track of a chain of custody for each sample. A chain of custody is a document that stays with the evidence at all times. Among other information, contains signatures and identification of all the people involved in transport, storage and analysis of the evidence.

This makes it much more difficult for intentional tampering to occur, it also acts as a detailed record of the location of the evidence at all times for record keeping purposes. It increases the reliability of a forensic chemist's work and increases the strength of the evidence in court. A distinction is made between destructive and non-destructive analytical methods. Destructive methods involve taking a sample from the object of interest, and so injures the object. Most spectroscopic techniques fall into this category. By contrast, a non-destructive method conserves the integrity of the object, and is generally preferred by forensic examiners. For example, optical microscopy and microspectroscopy cannot injure the sample, so they are considered non-destructive techniques.

Examples

Figure 2: Ozone cracking in Natural rubbertubing

Polymers for example, can be attacked by aggressive chemicals, and if under load, then cracks will grow by the mechanism of stress corrosion cracking. Perhaps the oldest known example is the ozone cracking of rubbers, where traces of ozone in the atmosphere attack double bonds in the chains of the materials. Elastomers with double bonds in their chains include natural rubber, nitrile rubber, and styrene-butadiene rubber. They are all highly susceptible to ozone attack, and can cause problems like car fires (from rubber fuel lines) and tire blow-outs. Nowadays, anti-ozonants are widely added to these polymers, so the incidence of cracking has dropped. However, not all safety-critical rubber products are protected, and, since it only takes a few parts per billion of ozone to start attack, failures are still occurring.

Another highly reactive gas is chlorine, which will attack susceptible polymers such as acetal resin and polybutylene pipework. There have been many examples of such pipes and acetal fittings failing in properties in the USA as a result of chlorine-induced cracking. In essence, the gas attacks sensitive parts of the chain molecules (especially secondary, tertiary or allylic carbon atoms), oxidising the chains and ultimately causing chain cleavage. The root cause is traces of chlorine in the water supply, added for its antibacterial action, attack occurring even at parts per million traces of the dissolved gas.

Most step-growth polymers can suffer hydrolysis in the presence of water, often a reaction catalysed by acid or alkali. Nylon for example, will degrade and crack rapidly if exposed to strong acids, a phenomenon well known to those who accidentally spill acid onto their shirts or tights. Polycarbonate is susceptible to alkali hydrolysis, the reaction simply depolymerising the material. Polyesters are prone to degrade when treated with strong acids, and, in all these cases, care must be taken to dry the raw materials for processing at high temperatures to prevent the problem from occurring.

Many polymers are also attacked by UV radiation at vulnerable points in their chain structures. Thus, polypropylene suffers severe cracking in sunlight unless antioxidants are added. The point of attack occurs at the tertiary carbon atom present in every repeat unit, causing oxidation and finally chain breakage.

Microscopy

Microscopy is the technical field of using microscopes to view samples and objects that cannot be seen with the unaided eye (objects that are not within the resolution range of the normal eye). There are three well-known branches of microscopy, optical, electron, and scanning probe microscopy.

Figure 3: Scanning electron microscope image of pollen.

Optical and electron microscopy involve the diffraction, reflection, or refraction of electromagnetic radiation/electron beams interacting with the specimen, and the subsequent collection of this scattered radiation or another signal in order to create an image. This process may be carried out by wide-field irradiation of the sample (for example standard light microscopy and transmission electron microscopy) or by scanning of a fine beam over the sample (for example confocal laser scanning microscopy and scanning electron microscopy). Scanning probe microscopy involves the interaction of a scanning probe with the surface of the object of interest.

The development of microscopy revolutionised biology and remains an essential technique in the life and physical sciences.

Optical Microscopy

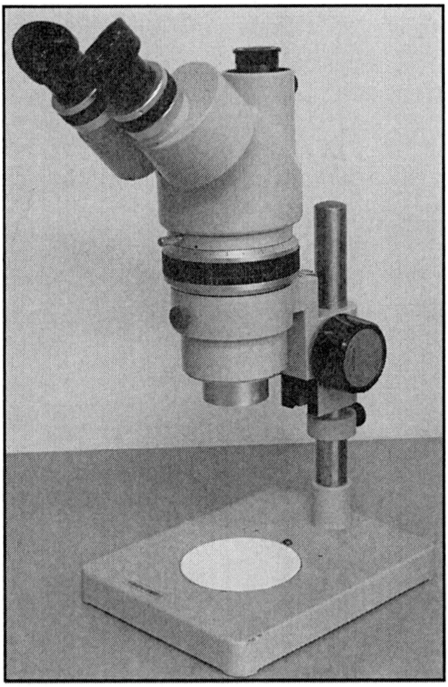

Figure 4: Stereo microscope

Optical or light microscopy involves passing visible light transmitted through or reflected from the sample through a single or multiple lenses to allow a magnified view of the sample. The resulting image can be detected directly by the eye, imaged on a photographic plate or captured digitally.

The single lens with its attachments, or the system of lenses and imaging equipment, along with the appropriate lighting equipment, sample stage and support, makes up the basic light microscope. The most recent development is the digital microscope, which uses a CCD camera to focus on the exhibit of interest. The image is shown on a computer screen, so eye-pieces are unnecessary.

Limitations

Limitations of standard optical microscopy (bright field microscopy) lie in three areas;

- The technique can only image dark or strongly refracting objects effectively.
- Diffraction limits resolution to approximately 0.2 micrometres.
- Out of focus light from points outside the focal plane reduces image clarity.

Live cells in particular generally lack sufficient contrast to be studied successfully, internal structures of the cell are colourless and transparent.

The most common way to increase contrast is to stain the different structures with selective dyes, but this involves killing and fixing the sample. Staining may also introduce artifacts, apparent structural details that are caused by the processing of the specimen and are thus not a legitimate feature of the specimen.

These limitations have all been overcome to some extent by specific microscopy techniques that can non-invasively increase the contrast of the image.

In general, these techniques make use of differences in the refractive index of cell structures. It is comparable to looking through a glass window: you (bright field microscopy) don't see the glass but merely the dirt on the glass.

There is however a difference as glass is a denser material, and this creates a difference in phase of the light passing through.

The human eye is not sensitive to this difference in phase but clever optical solutions have been thought out to change this difference in phase into a difference in amplitude (light intensity).

Techniques

In order to improve specimen contrast or highlight certain structures in a sample special techniques must be used.

A huge selection of microscopy techniques are available to increase contrast or label a sample.

Four examples of transilumination techniques used to generate contrast in a sample of tissue paper. 1.559 μm/pixel.

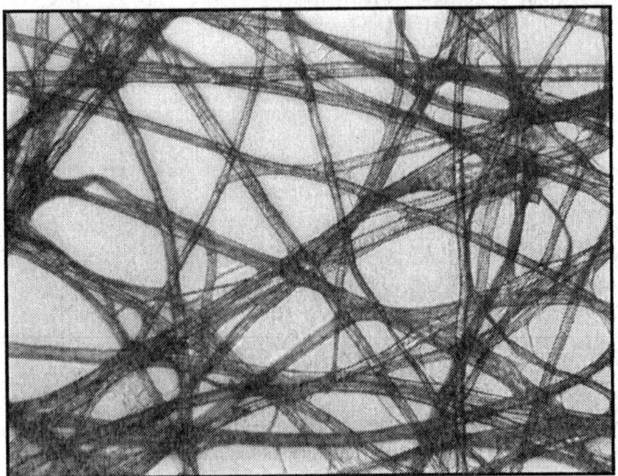

Figure 5: Bright field illumination, sample contrast comes from absorbance of light in the sample.

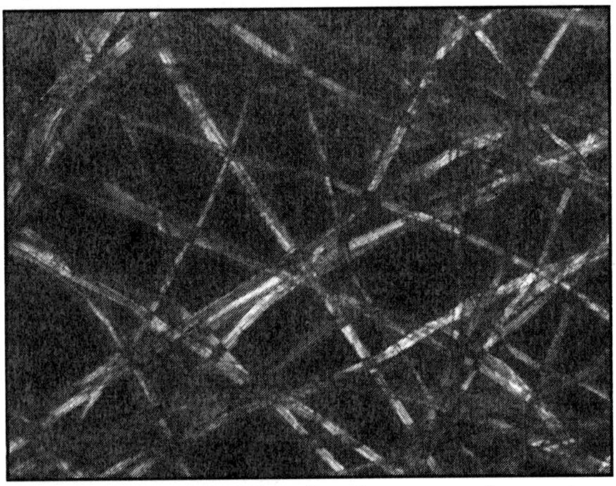

Figure 6: Cross-polarized lightillumination, sample contrast comes from rotation of polarized light through the sample.

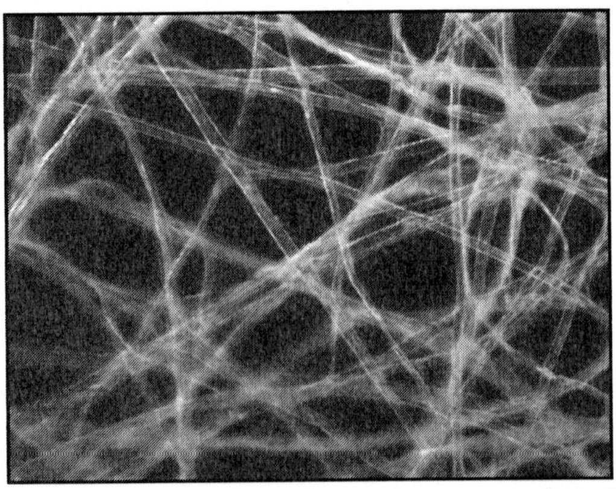

Figure 7: Dark field illumination, sample contrast comes from light scattered by the sample.

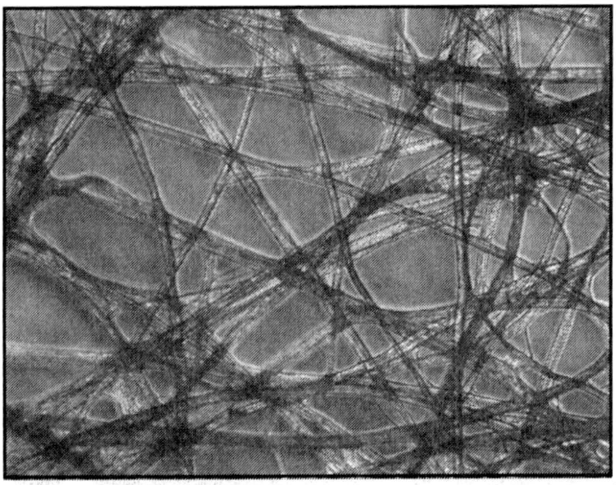

Figure 8: Phase contrastillumination, sample contrast comes frominterference of different path lengths of light through the sample.

Bright Field

Bright field microscopy is the simplest of all the light microscopy techniques. Sample illumination is via transmitted white light, i.e. illuminated from below and observed from above. Limitations include low contrast of most biological samples and low apparent resolution due to the blur of out of focus material. The simplicity of the technique and the minimal sample preparation required are significant advantages.

Oblique Lllumination

The use of oblique (from the side) illumination gives the image a 3-dimensional appearance and can highlight otherwise invisible features. A more recent technique based on this method is *Hoffmann's modulation contrast*, a system found on inverted microscopes for use in cell culture. Oblique illumination suffers from the same limitations as bright field microscopy (low contrast of many biological samples; low apparent resolution due to out of focus objects), but may highlight otherwise invisible structures.

Dark Field

Dark field microscopy is a technique for improving the contrast of unstained, transparent specimens. Dark field illumination uses a carefully aligned light source to minimise the quantity of directly-transmitted (unscattered) light entering the image plane, collecting only the light scattered by the sample. Darkfield can dramatically improve image contrast—especially of transparent objects – while requiring little equipment setup or sample preparation. However, the technique does suffer from low light intensity in final image of many biological samples, and continues to be affected by low apparent resolution.

Rheinberg illumination is a special variant of dark field illumination in which transparent, coloured filters are inserted just before the condenser so that light rays at high aperture are differently coloured than those at low aperture (i.e. the background to the specimen may be blue while the object appears self-luminous yellow). Other colour combinations are possible but their effectiveness is quite variable.

Dispersion Staining

Dispersion staining is an optical technique that results in a coloured image of a colourless object. This is an optical staining technique and requires no stains or dyes to produce a colour effect. There are five different microscope configurations used in the broader technique of dispersion staining. They include brightfield Becke' line, oblique, darkfield, phase contrast, and objective stop dispersion staining.

Phase Contrast

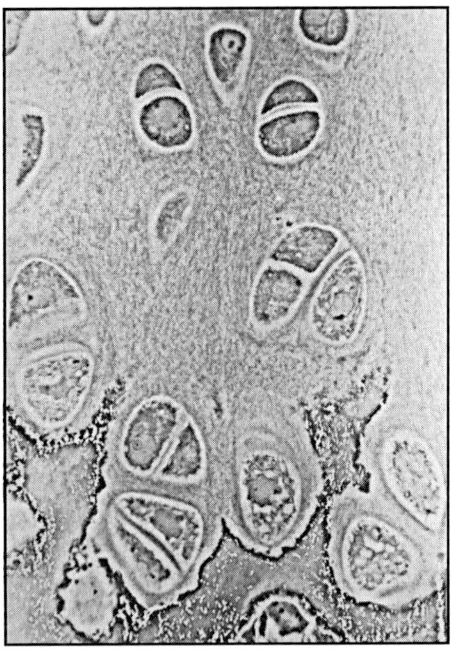

Figure 9: Phase-contrast image of uncalcified matrix and calcified matrix.

More sophisticated techniques will show proportional differences in optical density . Phase cóntrast is a widely used technique that shows differences in refractive index as difference in contrast. It was developed by the Dutch physicist Frits Zernike in the 1930s (for which he was awarded the Nobel Prize in 1953).

The nucleus in a cell for example will show up darkly against the surrounding cytoplasm. Contrast is excellent; however it is not for use with thick objects. Frequently, a halo is formed even around small objects, which obscures detail. The system consists of a circular annulus in the condenser, which produces a cone of light. This cone is superimposed on a similar sized ring within the phase-objective. Every objective has a different size ring, so for every objective another condenser setting has to be chosen. The ring in the objective has special optical properties: it first of all reduces the direct light in intensity, but more importantly, it creates an artificial phase difference of about a quarter wavelength.

Differential Interference Contrast

Superior and much more expensive is the use of interference contrast. Differences in optical density will show up as differences in relief. A nucleus within a cell will actually show up as a globule in the most often used differential interference contrast system according to Georges Nomarski. However, it has to be kept in mind that this is an *optical effect*, and the relief does not necessarily resemble the true shape. Contrast is very good and the condenser aperture can be used fully open, thereby reducing the depth of field and maximising resolution.

The system consists of a special prism (Nomarski prism, Wollaston prism) in the condenser that splits light in an ordinary and an extraordinary beam. The spatial difference between the two beams is minimal (less than the maximum resolution of the objective). After passage through the specimen, the beams are reunited by a similar prism in the objective.

In a homogeneous specimen, there is no difference between the two beams, and no contrast is being generated. However, near a refractive boundary (say a nucleus within the cytoplasm), the difference between the ordinary and the extraordinary beam will generate a relief in the image. Differential interference contrast requires a polarised light source to function; two polarising filters have to be fitted in the light path, one below the condenser (the polariser), and the other above the objective (the analyser).

Note: In cases where the optical design of a microscope produces an appreciable lateral separation of the two beams we have the case of classical interference microscopy, which does not result in relief images, but can nevertheless be used for the quantitative determination of mass-thicknesses of microscopic objects.

Interference Reflection Microscopy

An additional technique using interference is interference reflection microscopy (also known as reflected interference contrast, or RIC). It is used to examine the adhesion of cells to a glass surface, using polarised light of a narrow range of wavelengths to be reflected whenever there is an interface between two substances with different refractive indices. Whenever a cell is attached to the glass surface, reflected light from the glass and from the attached cell will interfere, while if there is no cell attached to the glass, there will be no interference.

Interference reflection microscopy can be obtained by using the same elements used by DIC, but without the prisms. Also, the light that is being detected is reflected and not transmitted as it is when DIC is employed.

Fluorescence

When certain compounds are illuminated with high energy light, they then emit light of a different, lower frequency. This effect is known as fluorescence. Often specimens show their own characteristic autofluorescence image, based on their chemical makeup.

This method is of critical importance in the modern life sciences, as it can be extremely sensitive, allowing the detection of single molecules. Many different fluorescent dyes can be used to stain different structures or chemical compounds. One particularly powerful method is the combination of antibodies coupled to a fluorophore as in immunostaining. Examples of commonly used fluorophores are fluorescein or rhodamine. The antibodies can be made tailored specifically for a chemical compound. For example, one strategy often in use is the artificial production of proteins, based on the genetic code (DNA). These proteins can then be used to immunize rabbits, which then form antibodies which bind to the protein. The antibodies are then coupled chemically to a fluorophore and then used to trace the proteins in the cells under study.

Highly-efficient fluorescent proteins such as the green fluorescent protein (GFP) have been developed using the molecular biology technique of gene fusion, a process that links the expression of the fluorescent compound to that of the target protein. This combined fluorescent protein is, in general, nontoxic to the organism and rarely interferes with the function of the protein under study. Genetically modified cells or organisms directly express the fluorescently-tagged proteins, which enables the study of the function of the original protein in vivo.

Growth of protein crystals results in both protein and salt crystals. Both are colourless and microscopic. Recovery of the protein crystals requires imaging which can be done by the intrinsic fluorescence of the protein or by using transmission microscopy. Both methods require a ultraviolet microscope as protein absorbs light at 280 nm. Protein will also fluorescence at approximately 353 nm when excited with 280 nm light.

Since fluorescence emission differs in wavelength (colour) from the excitation light, an ideal fluorescent image shows only the structure of interest that was labeled with the fluorescent dye. This high specificity led to the widespread use of fluorescence light microscopy in biomedical research. Different fluorescent dyes can be used to stain different biological structures, which can then be detected simultaneously, while still being specific due to the individual colour of the dye.

To block the excitation light from reaching the observer or the detector, filter sets of high quality are needed. These typically consist of an excitation filter selecting the range of excitation wavelengths, a dichroic mirror, and an emission filter blocking the excitation light. Most fluorescence microscopes are operated in the Epi-illumination mode (illumination and detection from one side of the sample) to further decrease the amount of excitation light entering the detector.

Confocal

Using a scanning point of light instead of full sample illumination confocal microscopy gives slightly higher resolution, and significant improvements in optical sectioning . Confocal microscopy is, therefore, commonly used where 3D structure is important.

Single Plane Illumination Microscopy and Light Sheet Florescence Microscopy

Using a plane of light formed by focusing light through a cylindrical lens at a narrow angle or by scanning a line of light in a plane perpendicular to the axis of objective, high resolution optical sections can be taken. Single plane illumination is also accomplished using beam shaping techniques incorporating multiple-prism beam expanders. The images are captured by CCDs. These variants allow very fast and high signal to noise ratio image capture.

Deconvolution

Fluorescence microscopy is extremely powerful due to its ability to show specifically labeled structures within a complex environment and also because of its inherent ability to provide three-dimensional information of biological structures. However, this information is blurred by the fact that, upon illumination, all fluorescently labeled structures emit light no matter whether they are in focus or not. This means that an image of a certain structure is always blurred by the contribution of light from structures that are out of focus. This phenomenon becomes apparent as a loss of contrast especially when using objectives with a high resolving power, typically oil immersion objectives with a high numerical aperture.

However, this phenomenon is not caused by random processes such as light scattering but can be relatively well defined by the optical properties of the image formation in the microscope imaging system. If one considers a small fluorescent light source (essentially a bright spot), light coming from this spot spreads out the further out of focus one is. Under ideal conditions, this produces a sort of "hourglass" shape of this point source in the third (axial) dimension. This shape is called the point spread function (PSF) of the microscope imaging system. Since any fluorescence image is made up of a large number of such small fluorescent light sources, the image is said to be "convolved by the point spread function".

Knowing this point spread function means that it is possible to reverse this process to a certain extent by computer-based methods commonly known as deconvolution microscopy. There are various algorithms available for 2D or 3D deconvolution. They can be roughly classified in *nonrestorative* and *restorative* methods. While the nonrestorative methods can improve contrast by removing out-of-focus light from focal planes, only the restorative methods can actually reassign light to its proper place of origin. This can be an advantage over other types of 3D microscopy such as confocal microscopy, because light is not thrown away but reused. For 3D deconvolution, one typically provides a series of images derived from different focal planes (called a Z-stack) plus the knowledge of the PSF, which can be derived either experimentally or theoretically from knowing all contributing parameters of the microscope.

Sub-diffraction Techniques

A multitude of super-resolution microscopy techniques has been developed in recent times which circumvent the diffraction barrier. This is mostly achieved by imaging a sufficiently static sample multiple times and either modifying the excitation light or observing stochastical changes in the image. Knowledge of and chemical control over fluorophore photophysics is at the core of these techniques, by which resolutions of ~20 nanometers are regularly obtained.

Serial Time-encoded Amplified Microscopy (STEAM)

Serial time encoded amplified microscopy (STEAM) is an imaging method that provides ultrafast shutter speed and frame rate, by using optical image amplification to circumvent the fundamental trade-off between sensitivity and speed, and a single-pixel photodetector to eliminate the need for a detector array and readout time limitations The method is at least 1000 times faster than the state of the art CCD and CMOS cameras. Consequently, it is potentially useful for a broad range of scientific, industrial, and biomedical applications that require high image acquisition rates, including real-time diagnosis and evaluation of shockwaves, microfluidics, MEMS, and laser surgery.

Extensions

Most modern instruments provide simple solutions for micro-photography and image recording electronically. However such capabilities are not always present and the more experienced microscopist will, in many cases, still prefer a hand drawn image rather than a photograph. This is because a microscopist with knowledge of the subject can accurately convert a three dimensional image into a precise two dimensional drawing . In a photograph or other image capture system however, only one thin plane is ever in good focus. The creation of careful and accurate micrographs requires a microscopical technique using a monocular eyepiece. It is essential that both eyes are open and that the eye that is not observing down the microscope is instead concentrated on a sheet of paper on the bench besides the microscope. With practice, and without moving the head or eyes, it is possible to accurately record the observed details by tracing round the observed shapes by simultaneously "seeing" the pencil point in the microscopical image.

Practicing this technique also establishes good general microscopical technique. It is always less tiring to observe with the microscope focused so that the image is seen at infinity and with both eyes open at all times.

Other Enhancements

Microspectroscopy:spectroscopy with a microscope.

X-ray

As resolution depends on the wavelength of the light. Electron microscopy has been developed since the 1930s that use electron beams instead of light. Because of the much smaller wavelength of the electron beam, resolution is far higher.

Though less common, X-ray microscopy has also been developed since the late 1940s. The resolution of X-ray microscopy lies between that of light microscopy and electron microscopy.

Electron Microscopy

Until the invention of sub-diffraction microscopy, the wavelength of the light limited the resolution of traditional microscopy to around 0.2 micrometers. In order to gain higher resolution, the use of an electron beam with a far smaller wavelength is used in electron microscopes.

- Transmission electron microscopy (TEM) is quite similar to the compound light microscope, by sending an electron beam through a very thin slice of the specimen. The resolution limit in 2005 was around 0.05 nanometer and has not increased appreciably since that time.
- Scanning electron microscopy (SEM) visualises details on the surfaces of cells and particles and gives a very nice 3D view. It gives results much like those of the stereo light microscope, and, akin to that, its most useful magnification is in the lower range than that of the transmission electron microscope.

Scanning Probe Microscopy

This is a sub-diffraction technique. Examples of scanning probe microscopes are the atomic force microscope (AFM), the Scanning tunnelling microscope and the photonic force microscope. All such methods imply a solid probe tip in the vicinity (near field) of an object, which is supposed to be almost flat.

Ultrasonic Force

Ultrasonic Force Microscopy (UFM) has been developed in order to improve the details and image contrast on “flat” areas of interest where the AFM images are limited in contrast. The combination of AFM-UFM allows a near field acoustic microscopic image to be generated. The AFM tip is used to detect the ultrasonic waves and overcomes the limitation of wavelength that occurs in acoustic microscopy. By using the elastic changes under the AFM tip, an image of much greater detail than the AFM topography can be generated.

Ultrasonic force microscopy allows the local mapping of elasticity in atomic force microscopy by the application of ultrasonic vibration to the cantilever or sample. In an attempt to analyse the results of ultrasonic force microscopy in a quantitative fashion, a force-distance curve measurement is done with ultrasonic vibration applied to the cantilever base, and the results are compared with a model of the cantilever dynamics and tip-sample interaction based on the finite-difference technique.

Ultraviolet Microscopy

Ultraviolet microscopes have two main purposes. The first is to utilise the shorter wavelength of ultraviolet electromagnetic energy in order to improve the image resolution beyond that of the diffraction limit of standard optical microscopes. This technique is used for non-destructive inspection of devices with very small features such as those found in modern semiconductors. The second application for UV microscopes is contrast enhancement where the response of individual samples is enhanced, relative to their surrounding, due to the interaction of light with the molecules within the sample itself. One example is in the growth of protein crystals. Protein crystals are formed in salt solutions. As both salt and protein crystals are both formed in the growth process, and both are commonly transparent to the human eye, they cannot be differentiated with a standard optical microscope. As the tryptophan of protein absorbs light at 280 nm, whereas salt does not, imaging with a UV microscope with 280 nm bandpass filters makes it simple to differentiate between the two types of crystals as the protein crystals appear dark while the salt crystals are transparent.

Infrared Microscopy

The term infrared microscope *covers two main types of diffraction-limited microscopy. The first provides optical visualisation plus IR spectroscopic data collection. The second (more recent and more advanced) technique employs* focal plane array detection *for infrared chemical imaging, where the image contrast is determined by the response of individual sample regions to particular IR wavelengths selected by the user*. IR versions of sub-diffraction microscopy exist also. These include IR NSOM and photothermal microspectroscopy.

Digital Holographic Microscopy

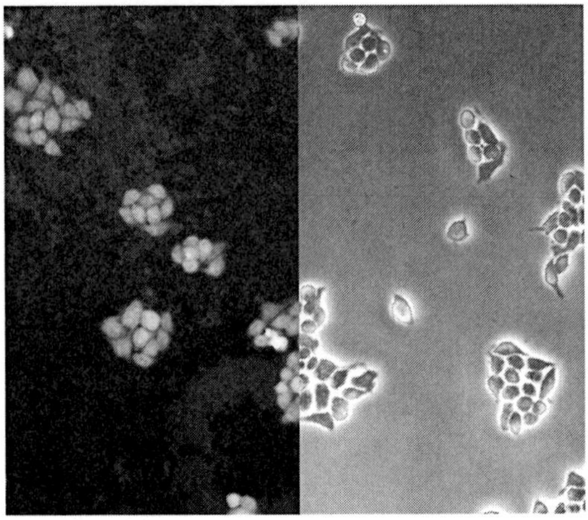

Figure 10: Human cells imaged by DHM phase shift (left) and phase contrast microscopy (right).

In digital holographic microscopy (DHM), interfering wave fronts from a coherent (monochromatic) light-source are recorded on a sensor. The image is digitally reconstructed by a computer from the recorded hologram. Besides the ordinary bright field image, a phase shift image is created as well. DHM can operate both in reflection and transmission mode. In reflection mode, the phase shift image provides a relative distance measurement and thus represents a topography map of the reflecting surface. In transmission mode, the phase shift image provides a label-free quantitative measurement of the optical thickness of the specimen. Phase shift images of biological cells are very similar to images of stained cells and have successfully been analysed by high content analysis software.

A unique feature of DHM is the ability to adjust focus after the image is recorded, since all focus planes are recorded simultaneously by the hologram. This feature makes it possible to image moving particles in a volume or to rapidly scan a surface. Another attractive feature is DHM's ability to use low cost optics by correcting optical aberrations by software.

Digital Pathology (Virtual Microscopy)

Digital Pathology is an image-based information environment enabled by computer technology that allows for the management of information generated from a digital slide. Digital pathology is enabled in part by virtual microscopy, which is the practice of converting glass slides into digital slides that can be viewed, managed, and analysed.

Laser Microscopy

Laser microscopy is a rapidly growing field that uses laser illumination sources in various forms of microscopy. For instance, laser microscopy focused on biological applications uses ultrashort pulse lasers, or femtosecond lasers, in a number of techniques labeled as nonlinear microscopy, saturation microscopy, and multiphoton fluorescence microscopy.

Amateur Microscopy

Amateur Microscopy *is the investigation and observation of biological and non-biological specimens for recreational purposes. Collectors of minerals, insects, seashells, and plants may use microscopes as tools to uncover features that help them classify their collected items. Other amateurs may be interested in observing the life found in pond water and of other samples. Microscopes may also prove useful for the water quality assessment for people that keep a home aquarium. Photographic documentation and drawing of the microscopic images are additional tasks that augment the spectrum of tasks of the amateur. There are even competitions for photomicrograph art. Participants of this pastime either may use commercially prepared microscopic slides or may engage in the task of specimen preparation.*

While microscopy is a central tool in the documentation of biological specimens, it is, in general, insufficient to justify the description of a new species based on microscopic investigations alone. Often genetic and biochemical tests are necessary to confirm the discovery of a new species. A laboratory and access to academic literature is a necessity, which is specialised and, in general, not available to amateurs. There is, however, one huge advantage that amateurs have above professionals: time to explore their surroundings. Often, advanced amateurs team up with professionals to validate their findings and (possibly) describe new species.

In the late 1800s, amateur microscopy became a popular hobby in the United States and Europe. Several 'professional amateurs' were being paid for their sampling trips and microscopic explorations by philanthropists, to keep them amused on the Sunday afternoon (e.g., the diatom specialist A. Grunow, being paid by (among others) a Belgian industrialist). Professor John Phin published "Practical Hints on the Selection and Use of the Microscope (Second Edition, 1878)," and was also the editor of the "American Journal of Microscopy."

In 1995, a loose group of amateur microscopists, drawn from several organisations in the UK and US, founded a site for microscopy based on the knowledge and input of amateur (perhaps better referred to as 'enthusiast') microscopists.

This was the first attempt to establish 'amateur' microscopy as a serious subjectin the then-emerging new media of the Internet. Today, it remains as a powerful established international resource for all ages, to input their findings and share information. It is a nonprofit-making web presence dedicated to the pursuit of science and understanding of the small-scale world.

Examples of amateur microscopy images:

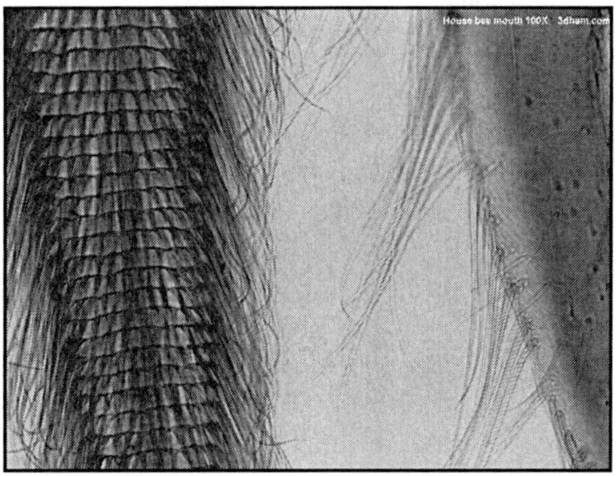

Figure 11: "house bee" Mouth 100X

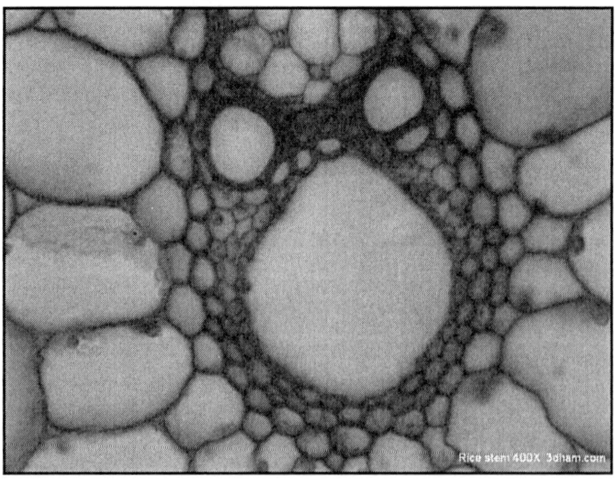

Figure 12: Rice Stem cs 400X

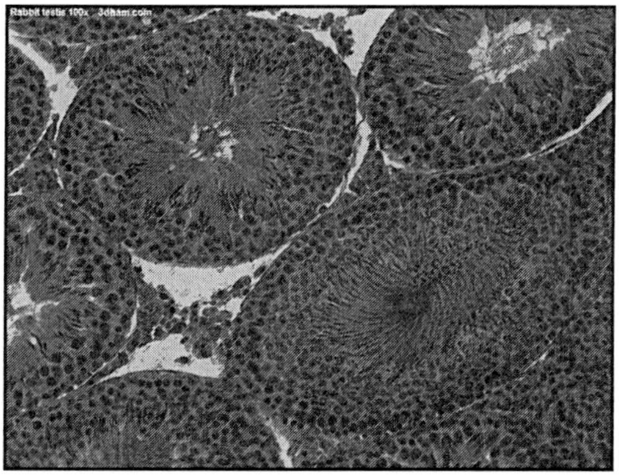

Figure 13: Rabbit Testis 100X

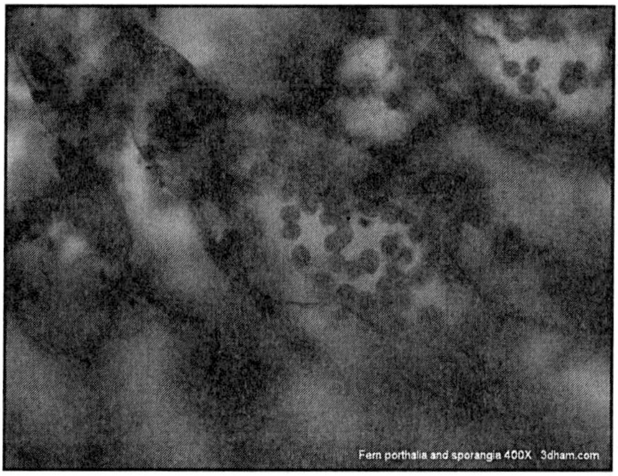

Figure 14: Fern Porothallium 400X

Digital Microscope

A digital microscope is a variation of a traditional optical microscope that uses optics and a charge-coupled device (CCD) camera to output a digital image to a monitor, sometimes by means of software running on a computer. A digital microscope differs from an optical microscope in that there is no provision to observe the sample directly through an eyepiece. Since the optical image is projected directly on the CCD camera, the entire system is designed for the monitor image. The optics for the human eye are omitted.

Digital microscope is the term commonly used for low price commercial microscopes designed for use with a computer. Although the term can also be used for advanced systems that do not require a separate computer and can cost tens of thousands of dollars. The low price commercial microscopes normally omit the optics for illumination (for example Köhler illumination and phase contrast illumination) and are more akin to webcams with a macro lens. For information about optical microscopes with a digital camera in research and development see optical microscope.

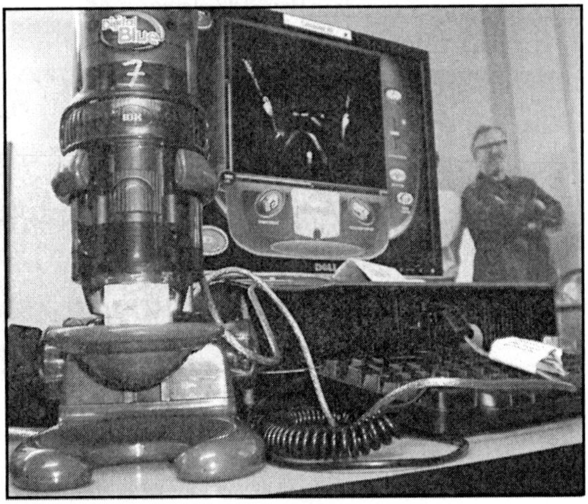

Figure 15: An insect observed with a digital microscope.

History

An early digital microscope was made by a lens company in Tokyo, Japan in 1986, which is now known as Hirox Co Ltd. It included a control box and a lens connected to a computer. The original connection to the computer was analog through an S-video connection. Over time that connection was changed to Firewire 800 to handle the large amount of digital information coming from the digital camera. But some of their current advanced versions are all-in-one and do not require a computer. Other versions of digital microscopes were later developed by Keyence Corp and Leica Microsystems. Invention of the USB port resulted in a multitude of USB microscopes ranging in quality and magnification. They continue to fall in price, especially compared with conventional optical microscopes.

Stereo and Digital Microscopes

A primary difference between an optical microscope and a digital microscope is the magnification. With an optical microscope the magnification is found by multiplying the lens magnification by the eyepiece magnification. Since the digital microscope does not have an eyepiece, the magnification cannot be found using this method. Instead the magnification for a digital microscope is found by how many times larger the sample is reproduced on the monitor. Therefore the magnification will depend on the size of the monitor. The average digital microscope system has a 15" monitor, would result in an average difference in magnification between an optical microscope and a digital microscope of about 60%. Thus the magnification number of an optical microscope is usually 60% larger than the magnification number of a digital microscope.

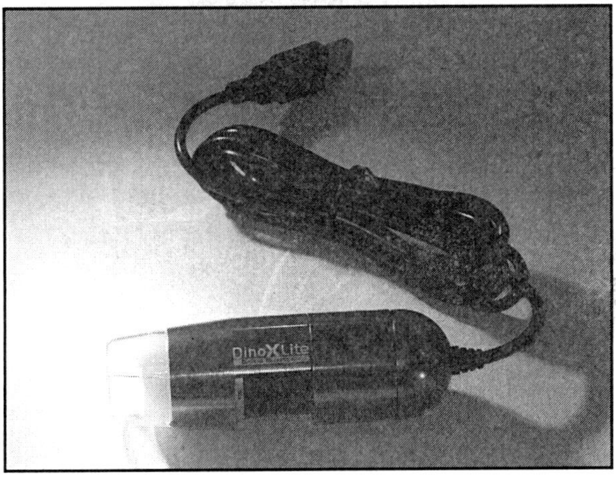

Figure 16: A miniature USB microscope.

Since the digital microscope has the image projected directly on to the CCD camera, it is possible to have higher quality recorded images than with an optical microscope. With the optical microscope, the lenses are made for the optics of the eye. Attaching a CCD camera to an optical microscope will result in a image that has compromises made for the eyepiece. Although the monitor image and recorded image may be of higher quality with the digital microscope, the application for the microscope may dictate which microscope is preferred.

Resolution

With a typical 2 megapixel CCD, a 1600×1200 pixels image is generated. The resolution of the image depends on the field of view of the lens used with the camera. The approximate pixel resolution can be determined by dividing the horizontalfield of view (FOV) by 1600. Increased resolution can be accomplished by creating a sub-pixel image. The Pixel Shift Method uses an actuator to physically move the CCD in order to take multiple overlapping images. By combining the images within the microscope, sub-pixel resolution can be generated. This method provides sub-pixel information, averaging a standard image is also a proven method to provide sub-pixel information.

2D Measurement

Most of the high end digital microscope systems have the ability to measure samples in 2D. The measurements are done onscreen by measuring the distance from pixel to pixel. This allows for length, width, diagonal, and circle measurements as well as much more. Some systems are even capable of counting particles.

3D Measurement

3D measurement is achieved with a digital microscope by image stacking. Using a step motor, the system takes images from the lowest focal plane in the field of view to the highest focal plane. Then it reconstructs theses images into a 3D model based on contrast to give a 3D colour image of the sample. From these 3D model measurements can be made, but their accuracy is based on the step motor and depth of field of the lens.

2D and 3D Tiling

2D and 3D tiling, also known as stitching or creating a panoramic, can now be done with the more advanced digital microscope systems. In 2D tiling the image is automatically tiled together seamlessly in real-time by moving the XY stage. 3D tiling combines the XY stage movement of 2D tiling with the Z-axis movement of 3D measurement to create a 3D panoramic.

Types

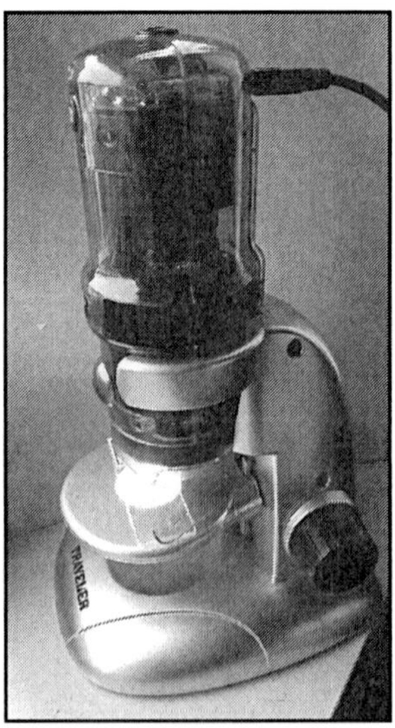

Figure 17: Cheap USB Digital Microscope

Digital microscopes range from inexpensive units costing from perhaps US$20 which connect to a computer via USB connector, to units costing tens of thousands of dollars. These advanced digital microscope systems usually are self contained, and do not require a computer.

Most of the cheaper microscopes which connect via USB have no stand, or a simple stand with clampable joints. They are essentially very simple webcams with small lenses and sensors—and can be used to view subjects which are not very close to the lens— mechanically arranged to allow focus at very close distances. Magnification is typically claimed to be user-adjustable from 10× to 200-400×.

Devices which connect to a computer require software to operate. Basic operation includes viewing the microscope image and recording "snapshots". More advanced functionality, possible even with simpler devices, includes recording moving images, time-lapse photography, measurement, image enhancement, annotation, etc. Many of the simpler units which connect to a computer use standard operating system facilities, and do not require device-specific drivers. A consequence of this is that many different microscope software packages can be used interchangeably with different microscopes, although such software may not support features unique to the more advanced devices. Basic operation may be possible with software included as part of computer operating systems—in Windows XP, images from microscopes which do not require special drivers can be viewed and recorded from "Scanners and Cameras" in Control Panel.

The more advanced digital microscope units have stands that hold the microscope and allow it to be racked up and down, similarly to standard optical microscopes. Calibrated movement in all three dimensions may be available. Resolution, image quality, and dynamic range vary with price. True spatial resolution may be increased by pixel shifting technology. Better units have larger sensors; the sensor may be cooled to reduce optical noise. Systems with a lower number of megapixels have a higher frame rate (30fps) and faster processing. The faster processing can be seen when using functions like HDR (high dynamic range). In addition to general-purpose microscopes, instruments specialised for specific applications are produced. These units can have a magnification range up to 0-7000x, are usually all-in-one (don't require a computer), and can be portable. They also differ from the cheaper USB microscopes in not only the quality of the image but the quality of the system's construction giving these types of systems a longer lifetime.

Chapter 4

Digital Pathology

Digital Pathology is an image-based information environment enabled by computer technology that allows for the management of information generated from a digital slide. Digital pathology is enabled in part by virtual microscopy, which is the practice of converting glass slides into digital slides that can be viewed, managed, and analysed.

Digital Pathology Environment

Scan: Digital slides are created from glass slides using a scanning device.

View: Digital slides are accessible for viewing via a computer monitor and viewing software.

Manage: Digital slides are maintained in an information management system that allows for archival and intelligent retrieval.

Analyse: Image analysis tools are used to derive objective quantification measures from digital slides.

Integrate: Digital pathology workflow is integrated into the institution's overall operational environment.

Sharing: Digital pathology also allows internet information sharing for education, diagnostics, publication and research.

Infrared Spectroscopy Correlation Table

In physical and analytical chemistry, infrared spectroscopy ("IR spectroscopy") is a technique used to identify chemical compounds based on how infrared radiation is absorbed by the compounds' chemical bonds. This article is an IR spectroscopy correlation table that lists some general absorption peaks for common types of atomic bonds and functional groups. Note that this information is unreferenced, and may not be accurate. The absorptions in this range do not apply only to bonds in organic molecules. IR spectroscopy is useful when it comes to analysis of inorganic compounds (such as metal complexes or fluoromanganates) as well.

Phosphonates could be also characterised by two middle-sized bands between 2300 and 2400 cm-1.

Rotational Spectroscopy

Rotational spectroscopy or microwave spectroscopy studies the absorption and emission of electromagnetic radiation (typically in the microwave region of the electromagnetic spectrum) by molecules associated with a corresponding change in the rotational quantum number of the molecule. The use of microwaves in spectroscopy essentially became possible due to the development of microwave technology for RADAR during World War II. Rotational spectroscopy is practical only in the gas phase where the rotational motion is quantised. Insolids or liquids the rotational motion is usually quenched due to collisions.

Rotational spectrum from a molecule (to first order) requires that the molecule have a dipole moment and that there be a difference between its center of charge and its center of mass, or equivalently a separation between two unlike charges. It is this dipole moment that enables the electric field of the light (microwave) to exert a torque on the molecule, causing it to rotate more quickly (in excitation) or slowly (in de-excitation). Homonuclear diatomic molecules such as dioxygen (O_2), dihydrogen (H_2), etc. do not have a dipole moment and, hence, no purely rotational spectrum. However, electronic excitations can lead to asymmetric charge distributions and thus provide a net dipole moment to the molecule. Under such circumstances, these molecules will exhibit a rotational spectrum.

Among the diatomic molecules, carbon monoxide (CO) has one of the simplest rotational spectra. As for triatomic molecules, hydrogen cyanide (HCa–N) has a simple rotational spectrum for a linear molecule and hydrogen isocyanide (HN=C:) for a non-linear molecule. As the number of atoms increases, the spectrum becomes more complex, as lines, due to different transitions, start overlapping.

Understanding the Rotational Spectrum

In quantum mechanics the free rotation of a molecule is quantised, that is the rotational energy and the angular momentum can take only certain fixed values; what these values are is related simply to the moment of inertia, I, of the molecule. In general, for any molecule, there are three moments of inertia: I_A, I_B and I_C about three mutually orthogonal axes A, B, and C with the origin at the center of mass of the system. Linear molecules are a special case because one of the moments of inertia (I_A, the moment of inertia along molecular axis) is negligible (i.e. $I_A \ll I_B = I_C$).

The general convention is to define the axes such that the axis A has the smallest moment of inertia, the axis B axis is the next smallest and so on such that $I_A \le I_B \le I_C$. (A rarely used convention defines the A axis along a symmetric axis of the molecule, if one exists. To avoid confusion, this article uses the prior convention.) The particular pattern of energy levels (and, hence, of transitions in the rotational spectrum) for a molecule is determined by its symmetry. A convenient way to look at the molecules is to divide them into four different classes (based on the symmetry of their structure). These are linear molecules, symmetric tops, spherical tops, and asymmetric tops. (Equivalently, many authors use the word "rotor" instead of "top".)

Linear Molecules

For a linear molecule the moments of inertia are related by $I_A \ll I_B = I_C$ For most of the purposes, I_A is taken to be zero. For a linear molecule, the separation of lines in the rotational spectrum can be related directly to the moment of inertia of the molecule, and, for a molecule of known atomic masses, can be used to determine the bond lengths (structure) directly. For diatomic molecules, this process is trivial, and can be made from a single measurement of the rotational spectrum. For linear molecules with more atoms, rather more work is required, and it is necessary to measure molecules in which more than one isotope of each atom have been substituted (effectively this gives rise to a set of simultaneous equations that can be solved for the bond lengths).

Examples of linear molecules: dioxygen (O=O), carbon monoxide (Oa–C*), hydroxy radical (OH), carbon dioxide (O=C=O), hydrogen cyanide (HCa–N), carbonyl sulfide (O=C=S), chloroethyne (HCa–CCl), acetylene (HCa–CH)

Symmetric Tops

A symmetric top is a molecule in which two moments of inertia are the same. As a matter of historical convenience, spectroscopists divide molecules into two classes of symmetric tops, *Oblate symmetric tops* (saucer or disc shaped) with $I_A = I_B < I_C$ and *Prolate symmetric tops* (rugby football, or cigar shaped) with $I_A < I_B = I_C$. The spectra look rather different, and are instantly recognisable. As with linear molecules, the structure of *symmetric tops* (bond lengths and bond angles) can be deduced from their spectra.

Examples of symmetric tops:

Oblate: benzene (C_6H_6), cyclobutadiene (C_4H_4), ammonia (NH_3)

Prolate: chloromethane (CH_3Cl), propyne ($CH_3Ca{-}CH$)

Spherical Tops

A spherical top molecule can be considered as a special case of *symmetric tops* with equal moment of inertia about all three axes ($I_A = I_B = I_C$).

Examples of spherical tops: phosphorus tetramer (P_4), carbon tetrachloride (CCl_4), methane (CH_4), ammonium ion (NH_4^+), sulfur hexafluoride (SF_6)

Asymmetric Tops

An asymmetric top occurs if all three moments of inertia are different. Most large molecules are asymmetric tops, even when they have a high degree of symmetry. In general, for such molecules, a simple interpretation of the spectrum is not normally possible. However, molecules have spectra similar to those of a linear molecule or a symmetric top, in which case the molecular structure may also be similar to that of a linear molecule or a symmetric top. For the most general case, however, all that can be done is to fit the spectra to three different moments of inertia. If the molecular formula is known, then educated guesses can be made of the possible structure, and, from this guessed structure, the moments of inertia can be calculated. If the calculated moments of inertia agree well with the measured moments of inertia, then the structure can be said to have been determined. For this approach to determining molecular structure, isotopic substitution is invaluable.

Examples of asymmetric tops: anthracene ($C_{14}H_{10}$), water (H_2O), nitrogen dioxide (NO_2)

Structure of Rotational Spectra

Linear Molecules

These molecules have two degenerate modes of rotation ($I_B = I_C$, $I_A = 0$). Since we cannot distinguish between the two modes, we need only one rotational quantum number (J) to describe the rotational motion of the molecule.

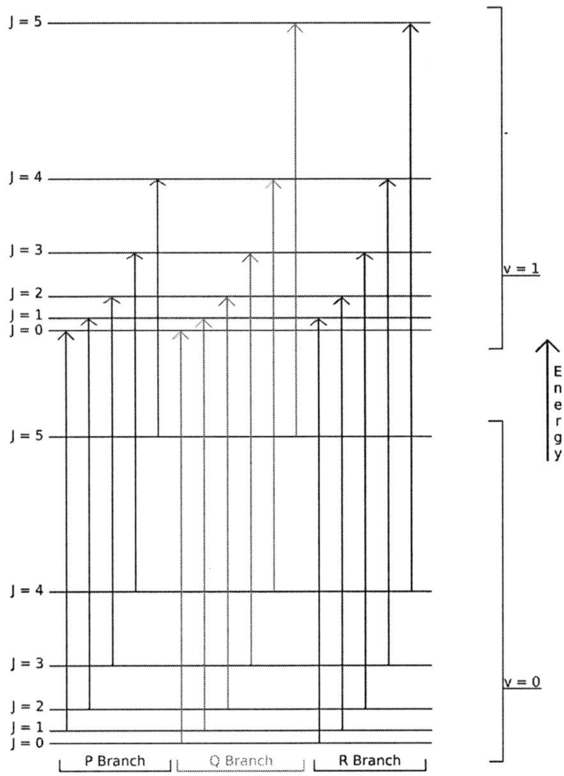

Figure 1: An energy level diagram showing some of the transitions involved in the IR vibration-rotation spectrum of a linear molecule: P branch (where $\Delta J = -1$), Q branch (not always allowed, $\Delta J = 0$) and R branch ($\Delta J = 1$)

The rotational energy levels ($F(J)$) of the molecule based on rigid rotor model can be expressed as,

$$F(J) = \tilde{B}_e J(J+1) \qquad J = 0,1,2,...$$

where $\tilde{B}_e$ is the rotational constant of the molecule and is related to the moment of inertia of the molecule $I_B = I_C$ by

$$\tilde{B}_e = \frac{h}{8\pi^2 c I_B}$$

(in wavenumber scale).

Selection rules dictate that during emission or absorption the rotational quantum number has to change by unity; i.e., $\Delta J = J' - J'' = \pm 1$ Thus, the locations of the lines in a rotational spectrum will be given by

$$\tilde{\nu}_{J' \leftrightarrow J''} = F\left(J'\right) - F\left(J''\right) =$$
$$2\tilde{B}_e\left(J'' + 1\right) \qquad J'' = 0,1,2,...$$

where J'' denotes the lower energy level and J' denotes higher energy level involved in the transition. The height of the lines is determined by the distribution of the molecules in the different levels and the probability of transition between two energy levels.

We observe that, for a rigid rotor, the transition lines are equally spaced in the wavenumber space. However, this is not always the case, except for the rigid rotor model. For non-rigid rotor model, we need to consider changes in the moment of inertia of the molecule. Two primary reasons for this are:

Centrifugal Distortion

When a molecule rotates, the centrifugal force pulls the atoms apart. As a result, the moment of inertia of the molecule increases, thus decreasing . To account for this a centrifugal distortion correction term is added to the rotational energy levels of the molecule.

$$F(J) = \tilde{B}_e J(J+1) - \tilde{D}_e J^2 (J+1)^2 \qquad J = 0,1,2,...$$

where $\tilde{D}_e$ is the centrifugal distortion constant.

Therefore, the line spacing for the rotational mode changes to,

$$\tilde{\nu}_{J' \leftrightarrow J''} = 2\tilde{B}_e\left(J'' + 1\right) - 4\tilde{D}_e\left(J'' + 1\right)^3 \qquad J'' = 0,1,2,...$$

Effect of Vibration on Rotation

A molecule is always in vibration. As the molecule vibrates, its moment of inertia changes. Further, there is a fictitious force, Coriolis coupling, between the vibrational motion of the nuclei in the rotating (non-inertial) frame. However, as long as the vibrational quantum number does not change (i.e., the molecule is in only one state of vibration), the effect of vibration on rotation is not important, because the time for vibration is much shorter than the time required for rotation. The Coriolis coupling is often negligible, too, if one is interested in low vibrational and rotational quantum numbers only.

Symmetric Top

The rotational motion of a symmetric top molecule can be described by two independent rotational quantum numbers (since two axes have equal moments of inertia, the rotational motion about these axes requires only one rotational quantum number for complete description). Instead of defining the two rotational quantum numbers for two independent axes, we associate one of the quantum number (J) with the total angular momentum of the molecule and the other quantum number (K) with the angular momentum of the axis that has different moment of inertia (i.e., axis C for oblate symmetric top and axis A for prolate symmetric tops). The rotational energy $F(J,K)$ of such a molecule, based on rigid rotor assumptions can be expressed in terms of the two previously defined rotational quantum numbers as follows

$$F(J,K) = \tilde{B}J(J+1) + (\tilde{A} - \tilde{B})K^2$$

$$J = 0,1,2,\ldots \quad \text{and} \quad K = -J, -J+1, \ldots, -1, 0, 1, \ldots, J-1, J$$

where $\tilde{B} = \dfrac{h}{8\pi^2 cI_B}$ and $\tilde{A} = \dfrac{h}{8\pi^2 cI_A}$ for a *prolate* symmetric top molecule or $\tilde{A} = \dfrac{h}{8\pi^2 cI_C}$ for an *oblate* molecule.

Selection rule for these molecules provide the guidelines for possible transitions. Therefore,

$$\Delta J = \pm 1 \quad \text{and} \quad \Delta K = 0$$

This is so because K is associated with the axis about which the molecule is symmetric and, hence, has no net dipole moment in that direction. Thus, there is no interaction of this mode with the light particles (photons).

This gives the transition wavenumbers of

$$\tilde{\nu}_{J' \leftrightarrow J'',K} = F(J',K) - F(J'',K) = 2\tilde{B}(J''+1) \qquad J'' = 0,1,\!$$

which is the same as in the case of a linear molecule.

In case of non-rigid rotors, the first order centrifugal distortion correction is given by

$$F(J,K) = \tilde{B}J(J+1) + (\tilde{A} - \tilde{B})K^2 - \tilde{D}_J J^2 (J+1)^2 - \tilde{D}_{JK} J(J+1)K^2 - D_K K^4 \qquad J = 0,1,2,\ldots \text{ an}$$

The suffixes on the centrifugal distortion constant D indicate the rotational mode involved and are not a function of the rotational quantum number. The location of the transition lines on a spectrum is given by

$$\tilde{\nu}_{J' \leftrightarrow J'',K} = F(J',K) - F(J'',K) = 2\tilde{B}(J''+1) - 4D_J(J''+1)^3 - 2D_{JK}(J''+1)K^2$$

Spherical Top

Unlike other molecules, spherical top molecules have no net dipole moment, and, hence, they do not exhibit a pure rotational spectrum.

Asymmetric Top

The spectrum for these molecules usually involves many lines due to three different rotational modes and their combinations. The following analysis is valid for the general case and collapses to the various special cases described above in the appropriate limit.

From the moments of inertia one can define an asymmetry parameter κ as $\kappa = \frac{2B - A - C}{A - C}$, which varies from 1 for a prolate symmetric top to 1 for an oblate symmetric top.

One can define a scaled rotational Hamiltonian dependent on J and κ. The (symmetric) matrix representation of this Hamiltonian is banded, zero everywhere but the main diagonal and the second subdiagonal. The Hamiltonian can be formulated in six different settings, dependent on the mapping of the principal axes to lab axes and handedness. For the most asymmetric, right-handed representation, the diagonal elements are $|k| \leq J$ for

$$H_{k,k}(\text{k}) = \text{k}k^2$$

and the second off-diagonal elements (independent of κ) are

$$H_{k,k+2}(\kappa) = \sqrt{[J(J+1)-(k+1)(k+2)][J(J+1)-k(k+1)]} \,/$$

Diagonalising H yields a set of $2J + 1$ scaled rotational energy levels E_k(κ). The rotational energy levels of the asymmetric rotor for total angular momentum J are then given by

$$\frac{A+C}{2} J(J+1) + \frac{A-C}{2} E_k(\kappa)$$

Hyperfine Interaction

In addition to the main structure that is observed in microwave spectra due to the rotational motion of the molecules, a whole host of further interactions are responsible for small details in the spectra, and the study of these details provides a very deep understanding of molecular quantum mechanics. The main interactions responsible for small changes in the spectra (additional splittings and shifts of lines) are due to magnetic and electrostatic interactions in the molecule. The particular strength of such interactions differs in different molecules, but, in general, the order of these effects (in decreasing significance) is:

1. electron spin - electron spin interaction (this occurs in molecules with two or more unpaired electrons, and is a magnetic-dipole / magnetic-dipole interaction)
2. electron spin - molecular rotation (the rotation of a molecule corresponds to a magnetic dipole, which interacts with the magnetic dipole moment of the electron)
3. electron spin - nuclear spin interaction (the interaction between the magnetic dipole moment of the electron and the magnetic dipole moment of the nuclei (if present)).
4. electric field gradient - nuclear electric quadrupole interaction (the interaction between the electric field gradient of the electron cloud of the molecule and the electric quadrupole moments of nuclei (if present)).
5. nuclear spin - nuclear spin interaction (nuclear magnetic moments interacting with one another).

These interactions give rise to the characteristic energy levels that are probed in "magnetic resonance" spectroscopy such as NMR and ESR, where they represent the "zero field splittings," which are always present.

Experimental Determination of the Spectrum

This section requires expansion. Part of the rotational-vibrational spectrum of methane (CH4) gas (from FTIR), showing the presence of P-, Q- and R- branches (purple, top) and a simulation in PGOPHER (black, bottom). Frequency is on the x-axis, and transmittance on the y-axis.

Fourier transform infrared (FTIR) spectroscopy can be used to experimentally study rotational spectra. Typical spectra at these wavelengths involve rovibrational excitation, i.e., excitation of both a vibrational and a rotational mode of a molecule.

In the past, microwave spectra were determined using a simple arrangement in which low-pressure gas was introduced to a section of waveguide between a microwave source (of variable frequency) and a microwave detector. The spectrum was obtained by sweeping the frequency of the source while detecting the intensity of the transmitted radiation. This experimental arrangement has a major difficulty related to the propagation of microwave radiation through waveguides. The physical size of the waveguide restricts the frequency of the radiation that can be transmitted through it. For a given waveguide size (such as X-band), there is a cutoff frequency, and microwave radiation with smaller frequencies (longer wavelengths) cannot be propagated through the waveguide. In addition, as the frequency is increased, additional modes of propagation become possible, which correspond to different velocities of the radiation propagating down the waveguide (this can be envisaged as the radiation bouncing down the guide, at different angles of reflection). The net result of these considerations is that each size of waveguide is useful only over a rather narrow range of frequencies and must be physically swapped out for a different size of waveguide once this frequency range is exceeded.

From 1980 onward, microwave spectra have often been obtained using Fourier Transform Microwave Spectroscopy - a technique developed by W. H. Flygare at the University of Illinois.

Within the last two years, a further development of Fourier Transform Microwave Spectroscopy has occurred, which may well introduce a new renaissance into microwave spectroscopy. This is the use of "Chirped Pulses" to provide an electromagnetic wave that has as its Fourier Transform a very wide range of microwave frequencies.

Applications

This section requires expansion. Microwave spectroscopy is commonly used in physical chemistry to determine the structure of small molecules (such as ozone, methanol, or water) with high precision. Other common techniques for determining molecular structure, such as X-ray crystallography do not work very well for some of these molecules (especially the gases) and are not as precise. However, microwave spectroscopy is not useful for determining the structures of large molecules such as proteins.

Modern microwave spectrometers have very high resolution. When hyperfine structure can be observed, the technique can also provide information on the electronic structures of molecules.

Microwave spectroscopy is one of the principal means by which the constituents of the universe are determined from the earth. It is particularly useful for detecting molecules in the interstellar medium (ISM). One of the early surprises in interstellar chemistry came with the discovery of the existence in the ISM of long-chain carbon molecules. It was in attempting to research such molecules in the laboratory that Harry Kroto was led to the laboratory of Rick Smalley andRobert

Curl, where it was possible to vapourise carbon under enormous energy conditions. This collaborative experiment led to the discovery of C60, buckminsterfullerene, which led to the award of the 1996 Nobel prize in chemistry to Kroto, Smalley and Curl.

Molecular Vibration

A molecular vibration occurs when atoms in a molecule are in periodic motion while the molecule as a whole has constant translational and rotational motion. The frequency of the periodic motion is known as a vibration frequency, and the typical frequencies of molecular vibrations range from less than 10^{12} to approximately 10^{14} Hz.

In general, a molecule with N atoms has $3N - 6$ normal modes of vibration, but a *linear* molecule has $3N - 5$ such modes, as rotation about its molecular axis cannot be observed. A diatomic molecule has one normal mode of vibration. The normal modes of vibration of polyatomic molecules are independent of each other but each normal mode will involve simultaneous vibrations of different parts of the molecule such as different chemical bonds.

A molecular vibration is excited when the molecule absorbs a quantum of energy, E, corresponding to the vibration's frequency, ν, according to the relation $E = h\nu$ (where h is Planck's constant). A fundamental vibration is excited when one such quantum of energy is absorbed by the molecule in its ground state. When two quanta are absorbed the first overtone is excited, and so on to higher overtones.

To a first approximation, the motion in a normal vibration can be described as a kind of simple harmonic motion. In this approximation, the vibrational energy is a quadratic function (parabola) with respect to the atomic displacements and the first overtone has twice the frequency of the fundamental. In reality, vibrations are anharmonic and the first overtone has a frequency that is slightly lower than twice that of the fundamental. Excitation of the higher overtones involves progressively less and less additional energy and eventually leads to dissociation of the molecule, as the potential energy of the molecule is more like a Morse potential.

The vibrational states of a molecule can be probed in a variety of ways. The most direct way is through infrared spectroscopy, as vibrational transitions typically require an amount of energy that corresponds to the infrared region of the spectrum. Raman spectroscopy, which typically uses visible light, can also be used to measure vibration frequencies directly. Vibrational excitation can occur in conjunction with electronic excitation (vibronic transition), giving vibrational fine structure to electronic transitions, particularly with molecules in the gas state.

Simultaneous excitation of a vibration and rotations gives rise to vibration-rotation spectra.

Vibrational Coordinates

The coordinate of a normal vibration is a combination of changes *in the positions of atoms in the molecule. When the vibration is excited the coordinate changes sinusoidally with a frequency v, the frequency of the vibration.*

Internal Coordinates

Internal coordinates are of the following types, illustrated with reference to the planar molecule ethylene,

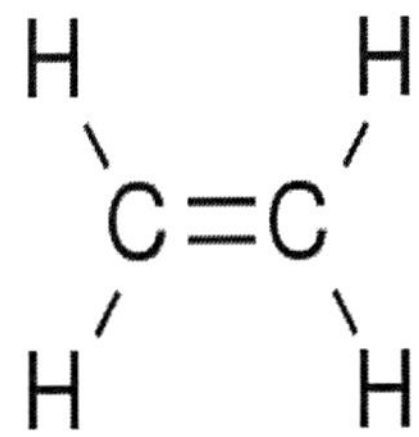

- Stretching: a change in the length of a bond, such as C-H or C-C
- Bending: a change in the angle between two bonds, such as the HCH angle in a methylene group
- Rocking: a change in angle between a group of atoms, such as a methylene group and the rest of the molecule.
- Wagging: a change in angle between the plane of a group of atoms, such as a methylene group and a plane through the rest of the molecule,
- Twisting: a change in the angle between the planes of two groups of atoms, such as a change in the angle between the two methylene groups.
- Out-of-plane: a change in the angle between any one of the C-H bonds and the plane defined by the remaining atoms of the ethylene molecule. Another example is in BF_3 when the boron atom moves in and out of the plane of the three fluorine atoms.

In a rocking, wagging or twisting coordinate the bond lengths within the groups involved do not change. The angles do. Rocking is distinguished from wagging by the fact that the atoms in the group stay in the same plane.

In ethene there are 12 internal coordinates: 4 C-H stretching, 1 C-C stretching, 2 H-C-H bending, 2 CH_2 rocking, 2 CH_2 wagging, 1 twisting. Note that the H-C-C angles cannot be used as internal coordinates as the angles at each carbon atom cannot all increase at the same time.

Symmetry-adapted Coordinates

Symmetry-adapted coordinates may be created by applying a projection operator to a set of internal coordinates. The projection operator is constructed with the aid of the character table of the molecular point group. For example, the four (unnormalised) C-H stretching coordinates of the molecule ethene are given by

$$Q_{s1} = q_1 + q_2 + q_3 + q_4$$
$$Q_{s2} = q_1 + q_2 - q_3 - q_4$$
$$Q_{s3} = q_1 - q_2 + q_3 - q_4$$
$$Q_{s4} = q_1 - q_2 - q_3 + q_4$$

where q_1 - q_4 are the internal coordinates for stretching of each of the four C-H bonds.

Illustrations of symmetry-adapted coordinates for most small molecules can be found in Nakamoto.

Normal Coordinates

The normal coordinates, denoted as Q, refer to the positions of atoms away from their equilibrium positions, with respect to a normal mode of vibration. Each normal mode is assigned a single normal coordinate, and so the normal coordinate refers to the "progress" along that normal mode at any given time. Formally, normal modes are determined by solving a secular determinant, and then the normal coordinates (over the normal modes) can be expressed as a summation over the cartesian cordinates (over the atom positions). The advantage of working in normal modes is that they diagonalize the matrix governing the molecular vibrations, so each normal mode is an independent molecular vibration, associated with its own spectrum of quantum mechanical states. If the molecule possesses symmetries, it will belong to a point group, and the normal modes will "transform as" an irreducible representation under that group. The normal modes can then be qualitatively determined by applying group theory and projecting the irreducible representation onto the cartesian coordinates. For example, when this treatment is applied to CO_2, it is found that the C=O stretches are not independent, but rather there is a O=C=O symmetric stretch and an O=C=O asymmetric stretch.

- symmetric stretching: the sum of the two C-O stretching coordinates; the two C-O bond lengths change by the same amount and the carbon atom is stationary. $Q = q_1 + q_2$
- asymmetric stretching: the difference of the two C-O stretching coordinates; one C-O bond length increases while the other decreases. $Q = q_1 - q_2$.

When two or more normal coordinates belong to the same irreducible representation of the molecular point group (colloquially, have the same symmetry) there is "mixing" and the coefficients of the combination cannot be determined *a priori*. For example, in the linear molecule hydrogen cyanide, HCN, The two stretching vibrations are

1. principally C-H stretching with a little C-N stretching; $Q_1 = q_1 + a\ q_2$ (a << 1)
2. principally C-N stretching with a little C-H stretching; $Q_2 = b\ q_1 + q_2$ (b << 1) The coefficients a and b are found by performing a full normal coordinate analysis by means of the Wilson GF method.

Newtonian Mechanics

Perhaps surprisingly, molecular vibrations can be treated using Newtonian mechanics to calculate the correct vibration frequencies. The basic assumption is that each vibration can be treated as though it corresponds to a spring. In the harmonic approximation the spring obeys Hooke's law: the force required to extend the spring is proportional to the extension. The proportionality constant is known as a force constant, k. *The anharmonic oscillator is considered elsewhere.*

$$\text{Force} = -kQ$$

By Newton's second law of motion this force is also equal to a reduced mass, μ, times acceleration.

$$\text{Force} = \mu\frac{d^2Q}{dt^2}$$

Since this is one and the same force the ordinary differential equation follows.

$$\mu\frac{d^2Q}{dt^2}+kQ=0$$

The solution to this equation of simple harmonic motion is

$$Q(t)=A\cos(2\pi\nu t);\ \nu=\frac{1}{2\pi}\sqrt{\frac{k}{\mu}}.$$

A is the maximum amplitude of the vibration coordinate Q. It remains to define the reduced mass, μ. In general, the reduced mass of a diatomic molecule, AB, is expressed in terms of the atomic masses, m_A and m_B, as

$$\frac{1}{\mu}=\frac{1}{m_A}+\frac{1}{m_B}.$$

The use of the reduced mass ensures that the centre of mass of the molecule is not affected by the vibration. In the harmonic approximation the potential energy of the molecule is a quadratic function of the normal coordinate. It follows that the force-constant is equal to the second derivative of the potential energy.

$$k=\frac{\partial^2V}{\partial Q^2}$$

When two or more normal vibrations have the same symmetry a full normal coordinate analysis must be performed. The vibration frequencies, V_i are obtained from the eigenvalues, λ_i, of the matrix product GF. G is a matrix of numbers derived from the masses of the atoms and the geometry of the molecule. F is a matrix derived from force-constant values. Details concerning the determination of the eigenvalues can be found in.

Quantum Mechanics

In the harmonic approximation the potential energy is a quadratic function of the normal coordinates. Solving the Schrödinger wave equation, the energy states for each normal coordinate are given by

$$E_n=h\nu=h\left(n+\frac{1}{2}\right)\frac{1}{2\pi}\sqrt{\frac{k}{m}}$$

where n is a quantum number that can take values of 0, 1, 2 ... The difference in energy when n changes by 1 are therefore equal to the energy derived using classical mechanics. Knowing the wave functions, certain selection rules can be formulated. For example, for a harmonic oscillator transitions are allowed only when the quantum number n changes by one,

$$\Delta n = \pm 1$$

but this does not apply to an anharmonic oscillator; the observation of overtones is only possible because vibrations are anharmonic. Another consequence of anharmonicity is that transitions such as between states n=2 and n=1 have slightly less energy than transitions between the ground state and first excited state. Such a transition gives rise to a hot band.

Intensities

In an infrared spectrum the intensity of an absorption band is proportional to the derivative of the molecular dipole moment with respect to the normal coordinate. The intensity of Raman bands depends on polarisability.

Near-infrared Spectroscopy

Near-infrared spectroscopy (NIRS) is a spectroscopic method that uses the near-infrared region of the electromagnetic spectrum (from about 800 nm to 2500 nm). Typical applications include pharmaceutical, medical diagnostics (including blood sugar and oximetry), food and agrochemical quality control, and combustion research, as well as cognitive neuroscience research.

Theory

Near-infrared spectroscopy is based on molecular overtone and combination vibrations. Such transitions are forbidden by the selection rules of quantum mechanics. As a result, the molar absorptivity in the near IR region is typically quite small. One advantage is that NIR can typically penetrate much farther into a sample than mid infrared radiation. Near-infrared spectroscopy is, therefore, not a particularly sensitive technique, but it can be very useful in probing bulk material with little or no sample preparation.

The molecular overtone and combination bands seen in the near IR are typically very broad, leading to complex spectra; it can be difficult to assign specific features to specific chemical components. Multivariate (multiple variables) calibration techniques (e.g., principal components analysis, partial least squares, orartificial neural networks) are often employed to extract the desired chemical information. Careful development of a set of calibration samples and application of multivariate calibration techniques is essential for near-infrared analytical methods.

History

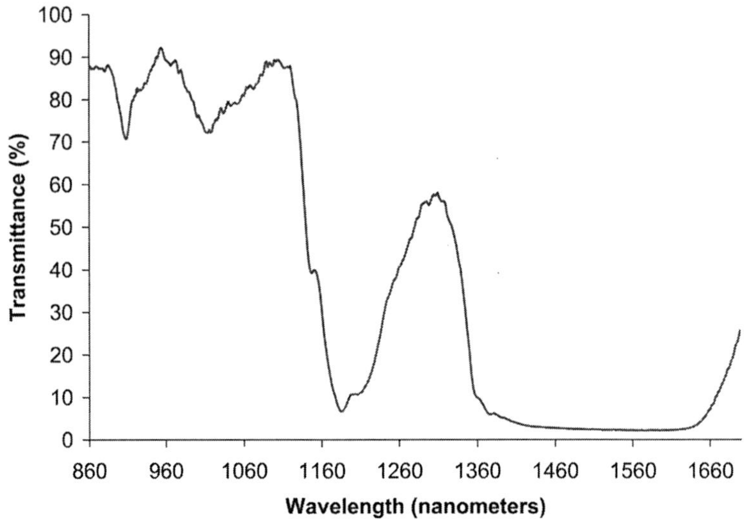

Figure 2: Near infrared spectrum of liquid ethanol.

The discovery of near-infrared energy is ascribed to Herschel in the 19th century, but the first industrial application began in the 1950s. In the first applications, NIRS was used only as an add-on unit to other optical devices that used other wavelengths such as ultraviolet (UV), visible (Vis), or mid-infrared (MIR) spectrometers. In the 1980s, a single unit, stand-alone NIRS system was made available, but the application of NIRS was focused more on chemical analysis. With the introduction of light-fibre optics in the mid-1980s and the monochromator-detector developments in early 1990s, NIRS became a more powerful tool for scientific research.

This optical method can be used in a number of fields of science including physics, physiology, or medicine. It is only in the last few decades that NIRS began to be used as a medical tool for monitoring patients.

Instrumentation

Instrumentation for near-IR (NIR) spectroscopy is similar to instruments for the UV-visible and mid-IR ranges. There is a source, a detector, and a dispersive element (such as a prism, or, more commonly, a diffraction grating) to allow the intensity at different wavelengths to be recorded. Fourier transform NIR instruments using an interferometer are also common, especially for wavelengths above

~1000 nm. Depending on the sample, the spectrum can be measured in either reflection or transmission. Common incandescent or quartz halogen light bulbs are most often used as broadband sources of near-infrared radiation for analytical applications. Light-emitting diodes (LEDs) are also used; they offer greater lifetime and spectral stability and reduced power requirements.

The type of detector used depends primarily on the range of wavelengths to be measured. Silicon-based CCDs are suitable for the shorter end of the NIR range, but are not sufficiently sensitive over most of the range (over 1000 nm). InGaAs and PbS devices are more suitable though less sensitive than CCDs. In certain diode array (DA) NIRS instruments, both silicon-based and InGaAs detectors are employed in the same instrument. Such instruments can record both UV-visible and NIR spectra 'simultaneously'.

Instruments intended for chemical imaging in the NIR may use a 2D array detector with a acousto-optic tunable filter. Multiple images may be recorded sequentially at different narrow wavelength bands.

Many commercial instruments for UV/vis spectroscopy are capable of recording spectra in the NIR range (to perhaps ~900 nm). In the same way, the range of some mid-IR instruments may extend into the NIR. In these instruments, the detector used for the NIR wavelengths is often the same detector used for the instrument's "main" range of interest.

Applications

Figure 3: NIR sensor for moisture measurement installed on a belt conveyor

The primary application of NIRS to the human body uses the fact that the transmission and absorption of NIR light in human body tissues contains information about hemoglobin concentration changes. When a specific area of the brain is activated, the localised blood volume in that area changes quickly. Optical imaging can measure the location and activity of specific regions of the brain by continuously monitoring blood hemoglobin levels through the determination of optical absorption coefficients. Typical applications of NIR spectroscopy include the analysis of foodstuffs, pharmaceuticals, combustion products and a major branch of astronomical spectroscopy.

Astronomical Spectroscopy

Near-infrared spectroscopy is used in astronomy for studying the atmospheres of cool stars where molecules can form. The vibrational and rotational signatures of molecules such as titanium oxide, cyanide, and carbon monoxide can be seen in this wavelength range and can give a clue towards the star's spectral type. It is also used for studying molecules in other astronomical contexts, such as in molecular clouds where new stars are formed. The astronomical phenomenon known as reddening means that near-infrared wavelengths are less affected by dust in the interstellar medium, such that regions inaccessible by optical spectroscopy can be studied in the near-infrared. Since dust and gas are strongly associated, these dusty regions are exactly those where infrared spectroscopy is most useful. The near-infrared spectra of very young stars provide important information about their ages and masses, which is important for understanding star formation in general.

Remote Monitoring

Techniques have been developed for NIR spectroscopic imaging. Hyperspectral imaging has been applied for a wide range of uses, including the remote investigation of plants and soils. Data can be collected from instruments on airplanes or satellites to assess ground cover and soil chemistry.

Materials Science

Techniques have been developed for NIR spectroscopy of microscopic sample areas for film thickness measurements, research into the optical characteristics of nanoparticles and optical coatings for the telecommunications industry.

Medical Uses

Medical applications of NIRS center on the non-invasive measurement of the amount and oxygen content of hemoglobin, as well as the use of exogenous optical tracers in conjunction with flow kinetics.

NIRS can be used for non-invasive assessment of brain function through the intact skull in human subjects by detecting changes in blood hemoglobin concentrations associated with neural activity, e.g., in branches of Cognitive psychology as a partial replacement for fMRI techniques. NIRS can be used on infants, and NIRS is much more portable than fMRI machines, even wireless instrumentation is available, which enables investigations in freely moving subjects . However, NIRS cannot fully replace fMRI because it can only be used to scan cortical tissue, where fMRI can be used to measure activation throughout the brain.

The application in functional mapping of the human cortex is called optical topography (OT), near infrared imaging (NIRI) or functional NIRS (fNIRS). The term optical tomography is used for three-dimensional NIRS. The terms NIRS, NIRI and OT are often used interchangeably, but they have some distinctions. The most important difference between NIRS and OT/NIRI is that OT/NIRI is used mainly to detect changes in optical properties of tissue simultaneously from multiple measurement points and display the results in the form of a map or image over a specific area, whereas NIRS provides quantitative data in absolute terms on up to a few specific points. The latter is also used to investigate other tissues such as, e.g., muscle, breast and tumours.

By employing several wavelengths and time resolved (frequency or time domain) and/or spatially resolved methods blood flow, volume and oxygenation can be quantified. These measurements are a form of oximetry. Applications of oximetry by NIRS methods include the detection of illnesses which affect the blood circulation (e.g., peripheral vascular disease), the detection and assessment of breast tumours, and the optimisation of training in sports medicine. These techniques can also be used for industry or agro processes in order to predict particle size/density.

The use of NIRS in conjunction with a bolus injection of indocyanine green (ICG) has been used to measure cerebral blood flow and cerebral metabolic rate of oxygen consumption in neonatal models.

NIRS is starting to be used in pediatric critical care, to help deal with cardiac surgery post-op. Indeed, NIRS is able to measure venous oxygen saturation (SVO_2), which is determined by the cardiac output, as well as other parameters (FiO_2, hemoglobin, oxygen uptake).

Therefore, following the NIRS gives critical care physicians a notion of the cardiac output. NIRS is liked by patients, because it is non-invasive, is painless, and uses non-ionizing radiation.

The instrumental development of NIRS/NIRI/OT has proceeded tremendously during the last years and, in particular, in terms of quantification, imaging and miniaturisation.

Particle Measurement

NIR is often used in particle sizing in a range of different fields, including studying pharmaceutical and agricultural powders.

Industrial Uses

As opposed to NIRS used in optical topography, general NIRS used in chemical assays does not provide imaging by mapping. For example, a clinical carbon dioxide analyser requires reference techniques and calibration routines to be able to get accurate CO_2 content change. In this case, calibration is performed by adjusting the zero control of the sample being tested after purposefully supplying 0% CO_2 or another known amount of CO_2 in the sample. Normal compressed gas from distributors contains about 95% O_2 and 5% CO_2, which can also be used to adjust %CO_2 meter reading to be exactly 5% at initial calibration.

Chapter 5

Resonance Raman Spectroscopy

Resonance Raman (RR) spectroscopy is a specialised implementation of the more general Raman spectroscopy.

As in Raman spectroscopy, RR spectroscopy provides information about the vibrations of molecules, and can also be used for identifying unknown substances. RR spectroscopy has found wide application to the analysis of bioinorganic molecules. Although the technique uses the same part of the electromagnetic spectrum as infrared (IR) spectroscopy, the two methods are actually complementary. Both are used to measure the energy required to change the vibrational state of a chemical compound.

IR spectroscopy involves measuring the direct absorption of photons with the appropriate energy to excite molecular bond vibrations. The wavelengths of these photons lie in the infrared region of the spectrum, hence the name of the technique. Raman spectroscopy measures the excitation of bond vibrations in an indirect manner. The two methods are complementary because some vibrational transitions that are observed in IR spectroscopy are not observed in Raman spectroscopy, and vice versa. RR spectroscopy is an improvement of traditional Raman spectroscopy that has increased sensitivity and is better suited for the study of complicated systems.

X-Ray Raman Scattering

In the x-ray region, enough energy is available for making electronic transitions possible. At core level resonances, X-Ray Raman Scattering can become the dominating part of the x-ray fluorescence spectrum. This is due to the resonant behavior of the Kramers-Heisenberg formula in which the denominator is minimised for incident energies that equal a core level. This type of scattering is also known as Resonant inelastic X-ray scattering (RIXS). In the soft x-ray range, RIXS has been shown to reflect crystal field excitations, which are often hard to observe with any other technique. Application of RIXS to strongly correlated materials is of particular value for gaining knowledge about their electronic structure. For certain wide band materials such as graphite, RIXS has been shown to (nearly) conserve crystal momentum and thus has found use as a complementary bandmapping technique.

History

Raman spectroscopy utilises the phenomenon of scattering. When light passes through a transparent sample, a fraction of the light is scattered in all directions. Most of the scattered photons are of the same wavelength of the incident light. This is known as Rayleigh scattering. However, in 1928, physicists C.V. Raman and K.S. Krishnan, and independently Grigory Landsberg and Leonid Mandelstam discovered that a small fraction of the scattered light had a different wavelength. Furthermore, this difference depended on the molecules present in the sample. For these observations, and his explanation of the phenomenon, Raman was awarded the 1930 Nobel Prize in physics. His explanation is now known as the theory of Raman scattering.

In the years following its discovery, Raman spectroscopy. was used to provide the first catalogue of molecular vibrational frequencies. Originally, heroic measures were required to obtain Raman spectra due to the low sensitivity of the technique. Typically, the sample was held in a long tube and illuminated along its length with a beam of filtered monochromatic light generated by a gas discharge lamp. The photons that were scattered by the sample were collected through an optical flat at the end of the tube. To maximise the sensitivity, the sample was highly concentrated (1 M or more) and relatively large volumes (5 mL or more) were used. Consequently, the use of Raman spectroscopy dwindled when commercial IRspectrophotometers became available in the 1940s. However, the advent of the laser in the 1960s resulted in simplified Raman spectroscopy instruments and also boosted the sensitivity of the technique. This has revived the use of Raman spectroscopy as a common analytical technique.

Theory of Raman Scattering

Degrees of Freedom

For any given chemical compound, there are a total of 3N degrees of freedom, where N is the number of atoms in the compound. This number arises from the ability of each atom in a molecule to move in three different directions (x, y, and z). When dealing with molecules, it is more common to consider the movement of the molecule as a whole. Consequently, the 3N degrees of freedom are partitioned into molecular translational, rotational, and vibrational motion. Three of the degrees of freedom correspond to translational motion of the molecule as a whole (along each of the three spatial dimensions). Similarly, three degrees of freedom correspond to rotations of the molecule about the x, y, and z-axes. However, linear molecules only have two rotations because rotations along the bond axis do not change the positions of the atoms in the molecule. The remaining degrees of freedom correspond to molecular vibrational modes. These modes include stretching and bending motions of the chemical bonds of the molecule. For a linear molecule, the number of vibrational modes is:

$$3N - 3 - 2 = 3N - 5$$

whereas for a non-linear molecule the number of vibrational modes are

$$3N - 6$$

Molecular Vibrations and Infrared Radiation

The frequencies of molecular vibrations range from less than 10^{12} to approximately 10^{14} Hz. These frequencies correspond to radiation in the infrared (IR) region of the electromagnetic spectrum. At any given instant, each molecule in a sample has a certain amount of vibrational energy. However, the amount of vibrational energy that a molecule has continually changes due to collisions and other interactions with other molecules in the sample.

At room temperature, most of molecules will be in the lowest energy state, which is known as the ground state. A few molecules will be in higher energy states, which are known as excited states. The fraction of molecules occupying a given vibrational mode at a given temperature can be calculated using the Boltzmann distribution. Performing such a calculation shows that, for relatively low temperatures (such as those used for most routine spectroscopy), most of the molecules occupy the ground vibrational state. Such a molecule can be excited to a higher vibrational mode through the direct absorption of a photon of the appropriate energy. This is the mechanism by which IR spectroscopy operates: infrared radiation is passed through the sample, and the intensity of the transmitted light is compared with that of the incident light. A reduction in intensity at a given wavelength of light indicates the absorption of energy by a vibrational transition. The energy, E, of a photon is

$$E = h\nu$$

where h is Planck's constant and ν is the frequency of the radiation. Thus, the energy required for such transition may be calculated if the frequency of the incident radiation is known.

Raman Scattering

It is also possible to observe molecular vibrations by an inelastic scattering process. In inelastic scattering, an absorbed photon is re-emitted with lower energy. In Raman scattering, the difference in energy between the absorbed and re-emitted photons corresponds to the energy required to excite a molecule to a higher vibrational mode.

Typically, in Raman spectroscopy high intensity laser radiation with wavelengths in either the visible or near-infrared regions of the spectrum is passed through a sample. Photons from the laser beam are absorbed by the molecules, exciting them to a virtual energy state. If the molecules relax back to the vibrational state that they started in, the reemitted photon has the same energy as the original photon. This leads to scattering of the laser light, but with no change in energy between the incoming photons and the reemitted/scattered photons. This type of scattering is known as Rayleigh scattering.

However, it is possible for the molecules to relax back to a vibrational state that is higher in energy than the state they started in. In this case, the original photon and the reemitted photon differ in energy by the amount required to vibrationally excite the molecule. Generally, the difference in energy is recorded as the difference in wavenumber ($\Delta\bar{\nu}$) between the laser light and the scattered light. A Raman spectrum is generated by plotting the intensity of the reemitted light versus $\Delta\bar{\nu}$ In this example the reemitted radiation is lower in energy than the incident laser light. Consequently, the change in wavenumber is positive and results in a series of peaks in the Raman spectrum known as Stokes lines.

A Raman spectrum also exhibits peaks that correspond to negative values of $\Delta\bar{\nu}$ These peaks are due to re-emitted photons that are higher in energy than the incident photons. This occurs when molecules that initially are in an excited vibrational state absorb the laser light and relax back to the lower vibrational state when they reemit the photon. These lines in the Raman spectrum are known as anti-Stokes lines. Since the Stokes lines and anti-Stokes lines gain and lose the same amount of energy, they are symmetric with respect to the peak due to elastic (Rayleigh) scattering ($\Delta\bar{\nu} = 0$). The anti-Stokes lines are appreciably less intense than the corresponding Stokes lines. This is because initially very few molecules are in excited vibrational states compared to the number in the ground state. Since anti-Stokes lines arise from the former and Stokes lines arise from the latter, the Stokes lines are much more intense. However, it should be noted that in molecules which exhibit fluorescence, the Stokes lines may be obscured while the anti-Stokes lines remain unaffected. In such cases, it is necessary to use the anti-Stokes lines despite their lower intensity.

Resonance Raman Spectroscopy

Raman spectroscopy can be used to identify chemical compounds because the values of $\Delta\bar{\nu}$ are indicative of different chemical species. This is because the frequencies of vibrational transitions depend on the atomic masses and the bond strengths. (Heavier atoms correspond to lower vibrational frequencies, while stronger bonds correspond to higher vibrational frequencies.) Thus, armed with a database of spectra from known compounds, one can unambiguously identify many different known chemical compounds based on a Raman spectrum. The number of vibrational modes scales with the number of atoms in a molecule, which means that the Raman spectra from large molecules will be very complicated. For example, proteins typically contain thousands of atoms and will therefore have thousands of vibrational modes. If these modes have similar energies ($\Delta\bar{\nu}$), then the spectrum may be incredibly cluttered and complicated.

Not all vibrational transitions will be "Raman active", i.e. some vibrational transitions will not appear in the Raman spectrum. This is because of the spectroscopic selection rules for Raman. As opposed to IR spectroscopy, where a transition can only be seen when that particular vibration causes a net change in dipole moment of the molecule, in Raman only transitions where the polarisability of the molecule changes can be observed. This is due to the fundamental difference in how IR and Raman spectroscopy access the vibrational transitions. In Raman spectroscopy, the incoming photon causes a momentary distortion of the electron distribution around a bond in a molecule, followed by reemission of the radiation as the bond returns to its normal state. This causes temporary polarisation of the bond, and an induced dipole that disappears upon relaxation. In a molecule with a center of symmetry, a change in dipole is accomplished by loss of the center of symmetry, while a change in polarisability is compatible with preservation of the center of symmetry. Thus, in a centrosymmetric molecule, asymmetrical stretching and bending will be IR active and

Raman inactive, while symmetrical stretching and bending will be Raman active and IR inactive. Hence, in a centrosymmetric molecule, IR and Raman spectroscopy are mutually exclusive. For molecules without a center of symmetry, each vibrational mode may be IR active, Raman active, both, or neither. Symmetrical stretches and bends, however, tend to be Raman active.

In resonance Raman spectroscopy, the energy of the incoming laser is adjusted such that it or the scattered light coincide with an electronic transition of the molecule or crystal. In most materials the incoming and outgoing electronic resonances are sufficiently broad that they can not be distinguished. So, rather than exciting the molecule to a virtual energy state, it is excited to near one of its excited electronic transitions. Since the energy of these transitions differ from one chemical species to the next, this technique did not become applicable until the advent of tunable lasers in the early 1970s. (Tunable lasers are those where the wavelength can be altered within a specific range.) When the frequency of the laser beam is tuned to be near an electronic transition (resonance), the vibrational modes associated with that particular transition exhibit a greatly increased Raman scattering intensity. This usually overwhelms Raman signals from all of the other transitions. For instance, resonance with a π-π^* transition enhances stretching modes of the π-bonds involved with the transition, while the other modes remain unaffected.

This aspect of Raman spectroscopy becomes especially useful for large biomolecules with chromophores embedded in their structure. In such chromophores, the charge-transfer (CT) transitions of the metal complex generally enhance metal-ligand stretching modes, as well as some of modes associated with the ligands alone. Hence, in a biomolecule such as hemoglobin, tuning the laser to near the charge-transfer electronic transition of the iron center results in a spectrum reflecting only the stretching and bending modes associated with the tetrapyrrole-iron group. Consequently, in a molecule with thousands of vibrational modes, RR spectroscopy allows us to look at relatively few vibrational modes at a time. This reduces the complexity of the spectrum and allows for easier identification of an unknown protein. Also, if a protein has more than one chromophore, different chromophores can be studied individually if their CT bands differ in energy. In addition to identifying compounds, RR spectroscopy can also supply structural identification about chromophores in some cases.

The main advantage of RR spectroscopy over traditional Raman spectroscopy is the large increase in intensity of the peaks in question (by as much as a factor of 10^6). This allows RR spectra to be generated with sample concentrations as low as 10^{-8} M. This is in stark contrast to conventional Raman spectra, which usually requires concentrations greater than 0.01 M. Also, as previously mentioned, RR spectra usually exhibit only a few peaks, and different peaks can be selected for by targeting specific electronic transitions. The main disadvantage of RR spectroscopy is the increased risk of fluorescence and photodegradation of the sample due to the increased energy of the incoming laser light. Both of these factors can be minimised by using an infrared laser instead of visible light for non resonant Raman scattering, but not in RR where the laser must be tuned to the specific resonance, unless electronic levels of lower energy are available for the system under investigation.

Instrumentation

In RR spectroscopy, the light source consists of a tunable laser, whose radiation lies in either the near-infrared, visible, or near-ultraviolet regions of the spectrum. In creating a sample handling system, RR spectroscopy offers an advantage over IR spectroscopy in that glass can be used for windows, lenses, and other optical components. Another advantage over IR spectroscopy is that whereas water absorbs strongly in the IR spectrum and may mask other signals, it only gives a weak signal in Raman spectroscopy. Therefore, water can easily be used as a solvent. Since lasers can be easily focused on small surface areas, the risk of sample heating and photodegradation is diminished, and the emitted radiation can be focused more efficiently. Typically, the sample is placed into a tube, which can then be spun to further decrease the sample's exposure to the laser light, further diminishing the threat of photodegradation. Gaseous, liquid, and solidsamples can all be analysed using RR spectroscopy. Gas and liquid samples can be put directly into the sample chamber whereas solid samples must first be ground into a powder. With gaseous and solid samples, Raman scattering may still be too weak to easily detect. For these samples, the sample holder is placed between two mirrors that reflect the laser beam multiple times through the sample.

Since scattered light leaves the sample in all directions the probes that carry the scattered light back to the detector in Raman spectroscopy may be placed at any angle. Usually, the detector probes are most placed at an angle of 135° to the path of the exiting laser light beam. Two other common arrangements position the detector probe at 90° or 180° with respect to the incident light. Detection angles greater than 90° are generally called back-scattering detectors because they are oriented in the same direction as the incident laser light so the radiation must scatter back to the probes. In transmitting the incident laser light to the sample and the scattered light back to the detector, fibre-optic cables may be used. Such cables can transmit light 100 m or more, thereby allowing the analysis of samples under relatively rough experimental/environmental conditions.

After the scattered radiation exits the sample, it is sent through a monochromator. Typical monochromators consist of a diffraction grating mounted on a rotating platform. A diffraction grating causes light dispersion. Rotating the grating controls which wavelengths of scattered radiation reach the exit slit leading to the detector. The detector itself is usually a charge-coupled device (CCD), which allows the entire spectrum to be recorded simultaneously. Consequently, multiple scans can be acquired in a short period of time, which can drastically increase the signal-to-noise ratio of the spectrum. Currently, Raman spectrometers are more expensive than more traditional dispersive instruments. As the cost of tunable lasers decrease, RR spectroscopy should see increased use, especially in the studies of metal-ligand vibrations, which reside in a region that is typically very difficult to study by other instrumental techniques. With the advent of near-infrared tunable lasers, particularly the Ti-sapphire laser (which has a range of ~700-1100 nm), Fourier Transform Resonance Raman Spectrometers may soon be commercially available. These would offer the multiplex and Jaquinot advantages of Fourier Transform (FT) techniques.

Raman Scattering

Raman scattering or the Raman effect is the inelastic scattering of a photon. It was discovered by Sir Chandrasekhara Venkata Raman and Kariamanickam Srinivasa Krishnan in liquids, and by Grigory Landsberg andLeonid Mandelstam in crystals.

When photons are scattered from an atom or molecule, most photons are elastically scattered (Rayleigh scattering), such that the scattered photons have the same kinetic energy (frequency) and wavelength as the incident photons. However, a small fraction of the scattered electrons (approximately 1 in 10 million photons) is scattered by an excitation, with the scattered photons having a frequency different from, and usually lower than, the frequency of the incident photons. In a gas, Raman scattering can occur with a change in transitional energy of a molecule. Chemists are concerned primarily with the transitional Raman effect.

The inelastic scattering of light was predicted by Adolf Smekal in 1923 (and in German-language literature it may be referred to as the Smekal-Raman effect). In 1922, Indian physicist C. V. Raman published his work on the "Molecular Diffraction of Light," the first of a series of investigations with his collaborators that ultimately led to his discovery (on 28 February 1928) of the radiation effect that bears his name. The Raman effect was first reported by C. V. Raman and K. S. Krishnan, and independently by Grigory Landsberg and Leonid Mandelstam, in 1928. Raman received the Nobel Prize in 1930 for his work on the scattering of light. In 1998 the Raman effect was designated an ACS National Historical Chemical Landmark in recognition of its significance as a tool for analysing the composition of liquids, gases, and solids.

Stokes and Anti-Stokes Scattering

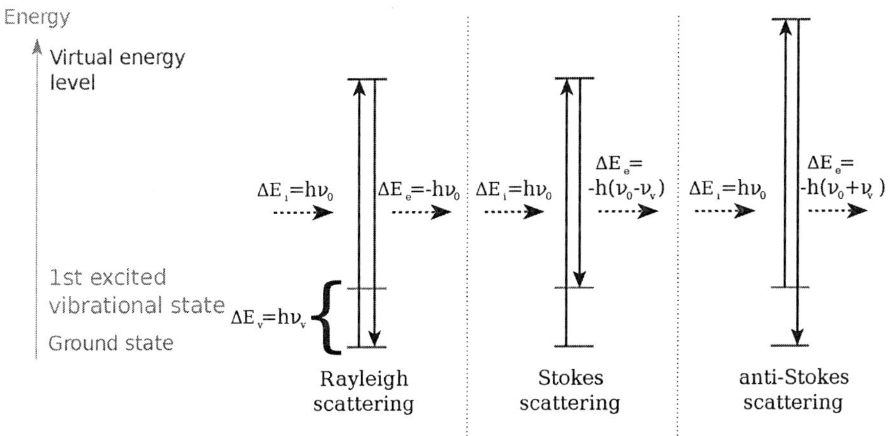

Figure 1: The different possibilities of visual light scattering: Rayleigh scattering (no Raman effect; the incident and emitted photons have the same energy), Stokes scattering (the atom or molecule absorbs energy; the emitted photon has less energy than the absorbed photon) and anti-Stokes scattering (the atom or molecule loses energy; the emitted photon has more energy than the absorbed photon)

There are two types of Raman scattering, Stokes scattering and anti-Stokes scattering.

The interaction of light with matter in a linear regime allows the absorption and emission of a photon precisely matching the difference in energy levels of the interacting electron or electrons.

The Raman effect corresponds, in perturbation theory, to the absorption and subsequent emission of a photon via an intermediate vibrational state, having a virtual energy level. There are three possibilities:

- no energy exchange between the incident photons and the molecules (and hence no Raman effect)
- energy exchanges occur between the incident photons and the molecules. The energy differences are equal to the differences of the vibrational and rotational energy-levels of the molecule. In crystals only specific phonons are allowed (solutions, which do not cancel themselves, of the wave equations) by the periodic structure, so Raman scattering can only appear at certain frequencies. In amorphous materials like glasses, more phononsare allowed and thereby the discrete spectral lines become broad.
- molecule absorbs energy: Stokes scattering. The resulting photon of lower energy generates a Stokes line on the red side of the incident spectrum.
- molecule loses energy: anti-Stokes scattering. Incident photons are shifted to the blue side of the spectrum, thus generating an anti-Stokes line.

These differences in energy are measured by subtracting the energy of the mono-energetic laser light from the energy of the scattered photons. The absolute value, however, doesn't depend on the process (Stokes or anti-Stokes scattering), because only the energy of the different vibrational levels is of importance. Therefore, the Raman spectrum is symmetric relative to the Rayleigh band. In addition, the intensities of the Raman bands are only dependent on the number of molecules occupying the different vibrational states, when the process began. If the sample is in thermal equilibrium, the relative numbers of molecules in states of different energy will be given by the Boltzmann distribution:

Thus lower energy states will have more molecules in them than will higher (excited) energy states. Therefore, the Stokes spectrum will be more intense than the anti-Stokes spectrum.

Distinction from Fluorescence

The Raman effect differs from the process of fluorescence. For the latter, the incident light is completely absorbed and the system is transferred to an excited state from which it can go to various lower states only after a certain resonance lifetime. The result of both processes is in essence the same: A photon with the frequency different from that of the incident photon is produced and the molecule is brought to a higher or lower energy level. But the major difference is that the Raman effect can take place for any frequency of the incident light. In contrast to the fluorescence effect, the Raman effect is therefore not a resonant effect. In practice, this means that a fluorescence peak is anchored at a specific excitation frequency, whereas a Raman peak maintains a constant separation from the excitation frequency. Another, related distinction is that Raman scattering is a coherent process, whereas fluorescence is not. This means that the measured intensity is the square of a coherent sum of scattering amplitudes. In practice, this means that different paths to the excitation of the same mode may interfere, leading to Fano effects: asymmetries in the shape of the scattering peaks.

Selection Rules

The distortion of a molecule in an electric field, and therefore the vibrational Raman cross section, is determined by its polarisability.

A Raman transition from one state to another, and therefore a Raman shift, can be activated optically only in the presence of non-zero polarisability derivative with respect to the normal coordinate (that is, the vibration or rotation):

$$\frac{\partial \alpha}{\partial Q} \neq 0$$

Raman-active vibrations/rotations can be identified by using almost any textbook that treats quantum mechanics or group theory for chemistry. Then, Raman-active modes can be found for molecules or crystals that show symmetry by using the appropriate character table for that symmetry group.

Stimulated Scattering and Amplification

Raman amplification can be obtained by using stimulated Raman scattering (SRS), which actually is a combination of a Raman process with stimulated emission. It is interesting for application in telecommunication fibers to amplify inside the standard material with low noise for the amplification process. However, the process requires significant power and thus imposes more stringent limits on the material. The amplification band can be up to 100 nm broad, depending on the availability of allowed photon states.

SRS is one of the processes that impede the laser coupling of inertial confinement fusion capsules.

Spectrum Generation

For high-intensity CW (continuous wave) lasers, SRS can be used to produce broad bandwidth spectra. This process can also be seen as a special case of four-wave mixing, wherein the frequencies of the two incident photons are equal and the emitted spectra are found in two bands separated from the incident light by the phonon energies. The initial Raman spectrum is built up with spontaneous emission and is amplified later on.

At high pumping levels in long fibers, higher-order Raman spectra can be generated by using the Raman spectrum as a new starting point, thereby building a chain of new spectra with decreasing amplitude. The disadvantage of intrinsic noise due to the initial spontaneous process can be overcome by seeding a spectrum at the beginning, or even using a feedback loop as in a resonator to stabilise the process. Since this technology easily fits into the fast evolving fibre laser field and there is demand for transversal coherent high-intensity light sources (i.e., broadband telecommunication, imaging applications), Raman amplification and spectrum generation might be widely used in the near-future.

Applications

Raman spectroscopy employs the Raman effect for materials analysis. The spectrum of the Raman scattered light depends on the molecular constituents present and their state, allowing the spectrum to be used for material identification and analysis. Raman spectroscopy is used to analyse a wide range of materials, including gases, liquids, and solids. Highly complex materials such as biological organisms and human tissue can also be analysed by Raman spectroscopy.

- For solid materials, Raman scattering is used as a tool to detect high-frequency phonon and magnon excitations.
- Raman lidar is used in atmospheric physics to measure the atmospheric extinction coefficient and the water vapour vertical distribution.
- Stimulated Raman transitions are also widely used for manipulating a trapped ion's energy levels, and thus basis qubit states.
- Raman amplification is used in optical amplifiers.

Nonlinear Optics

Nonlinear optics (NLO) is the branch of optics that describes the behaviour of light in *nonlinear media*, that is, media in which the dielectric polarisation P responds nonlinearly to the electric field E of the light. This nonlinearity is typically only observed at very high light intensities (values of the electric field comparable to interatomic electric fields, typically 10^8 V/m) such as those provided by pulsed lasers. In nonlinear optics, the superposition principle no longer holds.

Nonlinear optics remained unexplored until the discovery of Second harmonic generation shortly after demonstration of the first laser.

Nonlinear Optical Processes

Nonlinear optics gives rise to a host of optical phenomena:

Frequency Mixing Processes

- Second harmonic generation (SHG), or *frequency doubling*, generation of light with a doubled frequency (half the wavelength), two photons are destroyed creating a single photon at two times the frequency.
- Third harmonic generation (THG), generation of light with a tripled frequency (one-third the wavelength), three photons are destroyed creating a single photon at three times the frequency.
- High harmonic generation (HHG), generation of light with frequencies much greater than the original (typically 100 to 1000 times greater)
- Sum frequency generation (SFG), generation of light with a frequency that is the sum of two other frequencies (SHG is a special case of this)
- Difference frequency generation (DFG), generation of light with a frequency that is the difference between two other frequencies
- Optical parametric amplification (OPA), amplification of a signal input in the presence of a higher-frequency pump wave, at the same time generating an *idler* wave (can be considered as DFG)
- Optical parametric oscillation (OPO), generation of a signal and idler wave using a parametric amplifier in a resonator (with no signal input)

- Optical parametric generation (OPG), like parametric oscillation but without a resonator, using a very high gain instead
- Spontaneous parametric down conversion (SPDC), the amplification of the vacuum fluctuations in the low gain regime
- Optical rectification (OR), generation of quasi-static electric fields.
- Nonlinear light-matter interaction with free electrons and plasmas.

Other Nonlinear Processes

- Optical Kerr effect, intensity dependent refractive index (a $X^{(3)}$ effect)
- Self-focusing, an effect due to the Optical Kerr effect (and possibly higher order nonlinearities) caused by the spatial variation in the intensity creating a spatial variation in the refractive index
- Kerr-lens modelocking (KLM), the use of Self-focusing as a mechanism to mode lock laser.
- Self-phase modulation (SPM), an effect due to the Optical Kerr effect (and possibly higher order nonlinearities) caused by the temporal variation in the intensity creating a temporal variation in the refractive index
- Optical solitons, An equilibrium solution for either an optical pulse (temporal soliton) or Spatial mode (spatial soliton) that does not change during propagation due to a balance between diffraction and the Kerr effect (e.g. Self-phase modulation for temporal and Self-focusing for spatial solitons).
- Cross-phase modulation (XPM)
- Four-wave mixing (FWM), can also arise from other nonlinearities
- Cross-polarised wave generation (XPW), a $X^{(3)}$ effect in which a wave with polarisation vector perpendicular to the input one is generated
- Modulational instability
- Raman amplification
- Optical phase conjugation

- Stimulated Brillouin scattering, interaction of photons with acoustic phonons
- Multiphoton absorption, simultaneous absorption of two or more photons, transferring the energy to a single electron
- Multiple photoionization, near-simultaneous removal of many bound electrons by one photon
- Chaos in Optical Systems.

Related Processes

In these processes, the medium has a linear response to the light, but the properties of the medium are affected by other causes:

- Pockels effect, the refractive index is affected by a static electric field; used in electro-optic modulators;
- Acousto-optics, the refractive index is affected by acoustic waves (ultrasound); used in acousto-optic modulators.
- Raman scattering, interaction of photons with optical phonons;

Parametric Processes

Nonlinear effects fall into two qualitatively different categories, parametric and non-parametric effects. A parametric non-linearity is an interaction in which the quantum state of the nonlinear material is not changed by the interaction with the optical field. As a consequence of this, the process is 'instantaneous'; Energy and momentum conserving in the optical field, making phase matching important; and polarisation dependent.

Theory

Parametric and lossy 'instantaneous' (i.e. electronic) nonlinear optical phenomena, in which the optical fields are not too large, can be described by a Taylor series expansion of the dielectric Polarisation density (dipole moment per unit volume) *P(t)* at time *t* in terms of the electrical field:

$$P(t) \propto \chi^{(1)} E(t) + \chi^{(2)} E^2(t) + \chi^{(3)} E^3(t) + \cdots.$$

Here, the coefficients $X^{(n)}$ are the *n*-th order susceptibilities of the medium and the presence of such a term is generally referred to as an *n*-th order nonlinearity. In general X^n is an *n+1* order tensor representing both the polarisation dependent nature of the parametric interaction as well as the symmetries (or lack thereof) of the nonlinear material.

Wave-equation in a Nonlinear Material

Central to the study of electromagnetic waves is the wave equation. Starting with Maxwell's equations in an isotropic space containing no free charge, it can be shown that:

$$\nabla \times \nabla \times \boldsymbol{E} + \frac{n^2}{c^2}\frac{\partial^2}{\partial t^2}\boldsymbol{E} = -\frac{1}{c^2}\frac{\partial^2}{\partial t^2}\boldsymbol{P}^{NL},$$

where P^{NL} is the nonlinear part of the Polarisation density and n is the refractive index which comes from the linear term in P.

Note one can normally use the vector identity

$$\nabla \times (\nabla \times \boldsymbol{V}) = \nabla(\nabla \cdot \boldsymbol{V}) - \nabla^2 \boldsymbol{V}$$

and Gauss's law,

$$\nabla \cdot \boldsymbol{D} = 0,$$

to obtain the more familiar wave equation

$$\nabla^2 \boldsymbol{E} - \frac{n^2}{c^2}\frac{\partial^2}{\partial t^2}\boldsymbol{E} = 0.$$

For nonlinear medium Gauss's law does not imply that the identity

$$\nabla \cdot \boldsymbol{E} = 0$$

is true in general, even for an isotropic medium. However even when this term is not identically 0, it is often negligibly small and thus in practice is usually ignored giving us the standard nonlinear wave-equation:

$$\nabla^2 \boldsymbol{E} - \frac{n^2}{c^2}\frac{\partial^2}{\partial t^2}\boldsymbol{E} = \frac{1}{c^2}\frac{\partial^2}{\partial t^2}\boldsymbol{P}^{NL}.$$

Nonlinearities as a Wave Mixing Process

The nonlinear wave-equation is an inhomogeneous differential equation. The general solution comes from the study of Ordinary differential equations and can be solved by the use of a Green's function. Physically one gets the normalelectromagnetic wave solutions to the homogeneous part of the wave equation:

$$\nabla^2 \boldsymbol{E} - \frac{n^2}{c^2}\frac{\partial^2}{\partial t^2}\boldsymbol{E} = 0.$$

and the in homogenous term

$$\frac{1}{c^2}\frac{\partial^2}{\partial t^2}\boldsymbol{P}^{NL},$$

acts as a driver/source of the electromagnetic waves. One of the consequences of this is a nonlinear interaction will result in energy being mixed or coupled between different colours which is often called a 'wave mixing'. In general an n-th order will lead to $n+1$-th wave mixing. As an example, if we consider only a second order nonlinearity (three-wave mixing), then the polarisation, P, takes the form

$$P^{NL} = \mathrm{X}^{(2)}E^2(t).$$

If we assume that $E(t)$ is made up of two colours at frequencies ω_1 and ω_2, we can write $E(t)$ as

$$E(t) = E_1 e^{-i\omega_1 t} + E_2 e^{-i\omega_2 t} + c.c.$$

where $c.c.$ stands for complex conjugate. Plugging this into the expression for P gives

$$\begin{aligned} P^{NL} = \chi^{(2)} E^2(t) \;&= \chi^{(2)}[\,|E_1|^2\, e^{-i2\omega_1 t} + |E_2|^2\, e^{-i2\omega_2 t} \\ &+ 2E_1 E_2 e^{-i(\omega_1+\omega_2)t} \\ &+ 2E_1 E_2^* e^{-i(\omega_1-\omega_2)t} \\ &+ 2\left(|E_1| + |E_2|\right) e^0], \end{aligned}$$

which has frequency components at $2\omega_1, 2\omega_2$, $\omega_1+\omega_2$, $\omega_1-\omega_2$, and 0. These three-wave mixing processes correspond to the nonlinear effects known as Second harmonic generation, Sum frequency generation, Difference frequency generation and Optical rectification respectively.

Note: Parametric generation and amplification is a variation of difference frequency generation, where the lower-frequency of one of the two generating fields is much weaker (parametric amplification) or completely absent (parametric generation). In the latter case, the fundamental quantum-mechanical uncertainty in the electric field initiates the process.

Phase Matching

The above ignores the position dependence of the electrical fields. In a typical situation, the electrical fields are travelling waves described by

$$E_j(\boldsymbol{x},t) = e^{i(\omega_j t - \boldsymbol{k}_j \cdot \boldsymbol{x})},$$

at position x, with the wave vector $\boldsymbol{k}_j = n(\omega_j)\omega_j / c$, where c is the velocity of light and $n(\omega_j)$ the index of refraction of the medium at angular frequency ω_j.

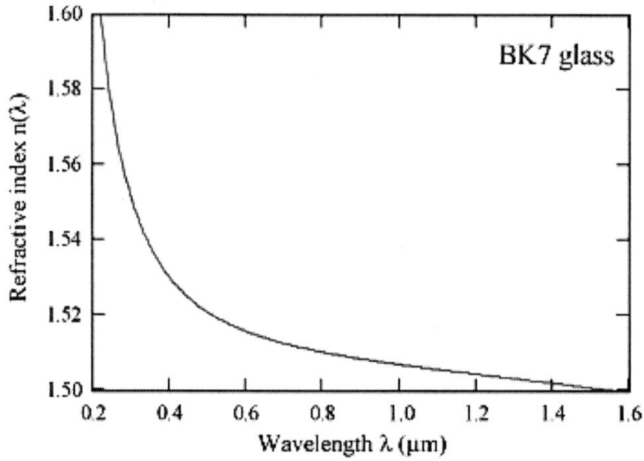

Figure 2: Most transparent materials, like the BK7 glass shown here, have normal dispersion: The index of refraction decreases monotonically as a function of wavelength (or increases as a function of frequency). This makes phase-matching impossible in most frequency-mixing processes. For example, in SHG, there is no simultaneous solution to ω' = 2ω and k'=2k in these materials. Birefringent materials avoid this problem by having two indices of refraction at once.

Thus, the second-order polarisation angular frequency ω_3 is

$$P^{(2)}(\boldsymbol{x},t) \propto E_1^{n_1} E_2^{n_2} e^{i(\omega_3 t-(m_1\boldsymbol{k}_1+m_2\boldsymbol{k}_2)\cdot\boldsymbol{x})}.$$

At each position , the oscillating second-order polarisation radiates at angular frequency ω_3 and a corresponding wave vector $\boldsymbol{k}_3 = n(\omega_3)\omega_3 / c$ Constructive interference, and therefore a high intensity ω_3 field, will occur only if

$$\boldsymbol{k}_3 = m_1\boldsymbol{k}_1 + m_2\boldsymbol{k}_2.$$

The above equation is known as the *phase matching condition.* Typically, three-wave mixing is done in a birefringent crystalline material (I.e., the refractive index depends on the polarisation and direction of the light that passes through.), where the polarisations of the fields and the orientation of the crystal are chosen such that the phase-matching condition is fulfilled. This phase matching technique is called angle tuning. Typically a crystal has three axes, one or two of which have a different refractive index than the other one(s). Uniaxial crystals, for example, have a single preferred axis, called the extraordinary (e) axis, while the other two are ordinary axes (o). There are several schemes of choosing the polarisations for this crystal type. If the signal and idler have the same polarisation, it is called "Type-I phase-matching", and if their polarisations are perpendicular, it is called "Type-II phase-matching". However, other conventions exist that specify further which frequency has what polarisation relative to the crystal axis. These types are listed below, with the convention that the signal wavelength is shorter than the idler wavelength.

Table 1: Phase-matching types

Polarisations			***Scheme***
Pump	***Signal***	***Idler***	
e	o	o	Type I
e	o	e	Type II (or IIA)
e	e	o	Type III (or IIB)
e	e	e	Type IV
o	o	o	Type V
o	o	e	Type VI (or IIB or IIIA)
o	e	o	Type VII (or IIA or IIIB)
o	e	e	Type VIII (or I)

Most common nonlinear crystals are negative uniaxial, which means that the *e* axis has a smaller refractive index than the *o* axes. In those crystals, type I and II phasematching are usually the most suitable schemes. In positive uniaxial crystals, types VII and VIII are more suitable. Types II and III are essentially equivalent, except that the names of signal and idler are swapped when the signal has a longer wavelength than the idler. For this reason, they are sometimes called IIA and IIB. The type numbers V–VIII are less common than I and II and variants.

One undesirable effect of angle tuning is that the optical frequencies involved do not propagate collinearly with each other. This is due to the fact that the extraordinary wave propagating through a birefringent crystal possesses a Poynting vector that is not parallel with the propagation vector.

This would lead to beam walkoff which limits the nonlinear optical conversion efficiency. Two other methods of phase matching avoids beam walkoff by forcing all frequencies to propagate at a 90 degree angle with respect to the optical axis of the crystal. These methods are called temperature tuning and quasi-phase-matching.

Temperature tuning is where the pump (laser) frequency polarisation is orthogonal to the signal and idler frequency polarisation. The birefringence in some crystals, in particular Lithium Niobate is highly temperature dependent. The crystal is controlled at a certain temperature to achieve phase matching conditions.

The other method is quasi-phase matching. In this method the frequencies involved are not constantly locked in phase with each other, instead the crystal axis is flipped at a regular interval Λ, typically 15 micrometres in length. Hence, these crystals are called periodically poled. This results in the polarisation response of the crystal to be shifted back in phase with the pump beam by reversing the nonlinear susceptibility. This allows net positive energy flow from the pump into the signal and idler frequencies. In this case, the crystal itself provides the additional wavevector $k=2\pi/\lambda$ (and hence momentum) to satisfy the phase matching condition. Quasi-phase matching can be expanded to chirped gratings to get more bandwidth and to shape an SHG pulse like it is done in a dazzler. SHG of a pump and Self-phase modulation (emulated by second order processes) of the signal and an optical parametric amplifier can be integrated monolithically.

Higher-order Frequency Mixing

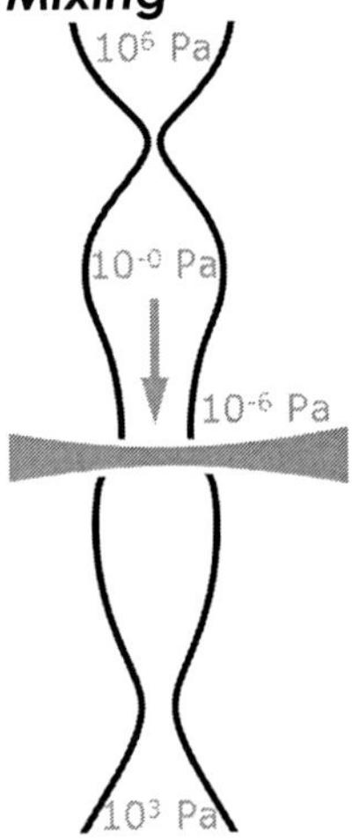

The above holds for $X^{(2)}$ processes. It can be extended for processes where $X^{(3)}$ is nonzero, something that is generally true in any medium without any symmetry restrictions. Third-harmonic generation is a $X^{(3)}$ process, although in laser applications, it is usually implemented as a two-stage process: first the fundamental laser frequency is doubled and then the doubled and the fundamental frequencies are added in a sum-frequency process. The Kerr effect can be described as a $X^{(3)}$ as well.

At high intensities the Taylor series, which led the domination of the lower orders, does not converge anymore and instead a time based model is used. When a noble gas atom is hit by an intense laser pulse, which has an electric field strength comparable to the Coulomb field of the atom, the outermost electron may be ionized from the atom. Once freed, the electron can be accelerated by the electric field of the light, first moving away from the ion, then back toward it as the field changes direction. The electron may then recombine with the ion, releasing its energy in the form of a photon. The light is emitted at every peak of the laser light field which is intense enough, producing a series of attosecond light flashes. The photon energies generated by this process can extend past the 800th harmonic order up to 1300 eV. This is called high-order harmonic generation. The laser must be linearly polarised, so that the electron returns to the vicinity of the parent ion. High-order harmonic generation has been observed in noble gas jets, cells, and gas-filled capillary waveguides.

Example Uses of Nonlinear Optics

Frequency Doubling

One of the most commonly used frequency-mixing processes is frequency doubling or second-harmonic generation. With this technique, the 1064-nm output from Nd:YAG lasers or the 800-nm output from Ti:sapphire lasers can be converted to visible light, with wavelengths of 532 nm (green) or 400 nm (violet), respectively.

Practically, frequency-doubling is carried out by placing a nonlinear medium in a laser beam. While there are many types of nonlinear media, the most common media are crystals. Commonly used crystals are BBO (**b**-barium borate), KDP (potassium dihydrogen phosphate), KTP (potassium titanyl phosphate), and lithium niobate. These crystals have the necessary properties of being strongly birefringent (necessary to obtain phase matching), having a specific crystal symmetry and of course being transparent for both the impinging laser light and the frequency doubled wavelength, and have high damage thresholds which make them resistant against the high-intensity laser light. However, organic polymeric materials are set to take over from crystals as they are cheaper to make, have lower drive voltages and superior performance.

Optical Phase Conjugation

It is possible, using nonlinear optical processes, to exactly reverse the propagation direction and phase variation of a beam of light. The reversed beam is called a *conjugate* beam, and thus the technique is known as optical phase conjugation (also called *time reversal, wavefront reversal* and *retroreflection*).

One can interpret this nonlinear optical interaction as being analogous to a real-time holographic process. In this case, the interacting beams simultaneously interact in a nonlinear optical material to form a dynamic hologram (two of the three input beams), or real-time diffraction pattern, in the material. The third incident beam diffracts off this dynamic hologram, and, in the process, reads out the phase-conjugate wave. In effect, all three incident beams interact (essentially) simultaneously to form several real-time holograms, resulting in a set of diffracted output waves that phase up as the "time-reversed" beam. In the language of nonlinear optics, the interacting beams result in a nonlinear polarisation within the material, which coherently radiates to form the phase-conjugate wave.

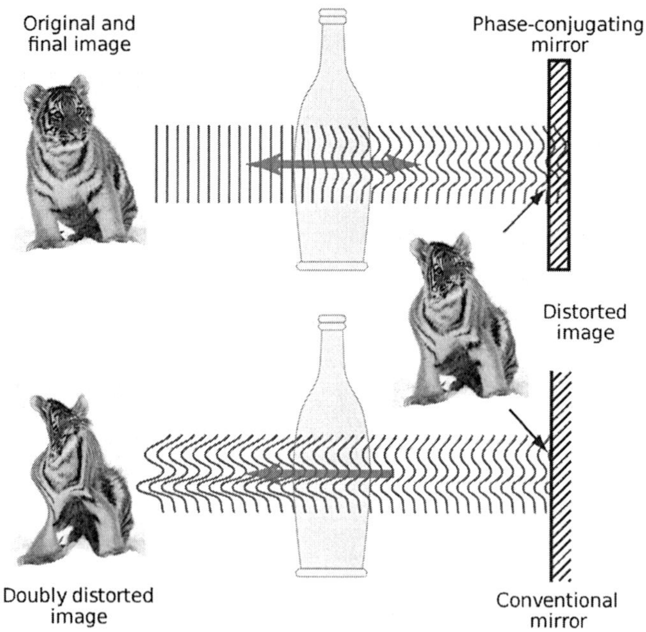

Figure 3: Comparison of a phase conjugate mirror with a conventional mirror. With the phase conjugate mirror the image is not deformed when passing through an aberrating element twice.

The most common way of producing optical phase conjugation is to use a four-wave mixing technique, though it is also possible to use processes such as stimulated Brillouin scattering. A device producing the phase conjugation effect is known as a phase conjugate mirror (PCM).

For the four-wave mixing technique, we can describe four beams (j = 1,2,3,4) with electric fields:

$$\Xi_j(\boldsymbol{x},t) = \frac{1}{2}E_j(\boldsymbol{x})e^{i(\omega_j t-\boldsymbol{k}\cdot\boldsymbol{x})} + \text{c.c.}$$

where E_j are the electric field amplitudes. Ξ_1 and Ξ_2 are known as the two pump waves, with Ξ_3 being the signal wave, and Ξ_4 being the generated conjugate wave.

If the pump waves and the signal wave are superimposed in a medium with a non-zero X(3), this produces a nonlinear polarisation field:

$$P_{\text{NL}} = \epsilon_0\chi^{(3)}(\Xi_1 + \Xi_2 + \Xi_3)^3$$

resulting in generation of waves with frequencies given by $\omega = \pm\omega_1 \pm\omega_2 \pm\omega_3$ in addition to third harmonic generation waves with $\omega = 3\omega_1$, $3\omega_2$, $3\omega_3$.

As above, the phase-matching condition determines which of these waves is the dominant. By choosing conditions such that $\omega = \omega_1 + \omega_2 - \omega_3$ and $k = k_1 + k_2 - k_3$, this gives a polarisation field:

$$P_\omega = \frac{1}{2}\chi^{(3)}\epsilon_0 E_1 E_2 E_3^* e^{i(\omega t-\boldsymbol{k}\cdot\boldsymbol{x})} + \text{c.c..}$$

This is the generating field for the phase conjugate beam, Ξ_4. Its direction is given by $k_4 = k_1 + k_2 - k_3$, and so if the two pump beams are counterpropagating ($k_1 = -k_2$), then the conjugate and signal beams propagate in opposite directions ($k_4 = -k_3$). This results in the retroreflecting property of the effect.

Further, it can be shown for a medium with refractive index n and a beam interaction length l, the electric field amplitude of the conjugate beam is approximated by

$$E_4 = \frac{i\omega l}{2nc}\chi^{(3)}E_1 E_2 E_3^*$$

(where c is the speed of light). If the pump beams E_1 and E_2 are plane (counterpropagating) waves, then:

$$E_4(\boldsymbol{x}) \propto E_3^*(\boldsymbol{x});$$

that is, the generated beam amplitude is the complex conjugate of the signal beam amplitude. Since the imaginary part of the amplitude contains the phase of the beam, this results in the reversal of phase property of the effect.

Note that the constant of proportionality between the signal and conjugate beams can be greater than 1. This is effectively a mirror with a reflection coefficient greater than 100%, producing an amplified reflection. The power for this comes from the two pump beams, which are depleted by the process.

The frequency of the conjugate wave can be different from that of the signal wave. If the pump waves are of frequency $\omega_1 = \omega_2 = \omega$, and the signal wave higher in frequency such that $\omega_3 = \omega + \Delta\omega$, then the conjugate wave is of frequency $\omega_4 = \omega - \Delta\omega$. This is known as *frequency flipping.*

Common SHG Materials

- 800 nm light : BBO
- 806 nm light : lithium iodate ($LiIO_3$)
- 860 nm light : potassium niobate ($KNbO_3$)
- 980 nm light : $KNbO_3$
- 1064 nm light : monopotassium phosphate (KH_2PO_4, KDP), lithium triborate (LBO) and **b**-barium borate (BBO).
- 1300 nm light : gallium selenide (GaSe)
- 1319 nm light : $KNbO_3$, BBO, KDP, potassium titanyl phosphate (KTP), lithium niobate ($LiNbO_3$), $LiIO_3$, and ammonium dihydrogen phosphate (ADP).

Parametric Process (Optics)

A parametric process is an optical process in which light interacts with matter in such a way as to leave the quantum state of the material unchanged. As a direct consequence of this there can be no transfer of energy, momentum, orangular momentum between the optical field and the physical system. In contrast a non-parametric process is a process in which any part of the quantum state of the system changes.

Temporal Characteristics

Since a parametric process prohibits a change in the energy state of the system, parametric processes are considered to be 'instantaneous' processes. This can be seen as follows; if an atom absorbs a photon with energy E, the atom's energy will increase by ΔE = E. Since we are assuming this is a parametric process, the quantum state cannot change and thus this energy state must be a Virtual state. By the Heisenberg Uncertainty Principle we know that ΔEΔt~', thus the lifetime of a parametric process is roughly Δt~'/ΔE, which is appreciably small for any non-zero ΔE.

Parametric Versus Non-parametric Processes

Linear Optics

In a linear optical system the dielectric polarisation, P, responds linearly to the presence of an electric field, E, and thus we can write

$$\boldsymbol{P} = \varepsilon_0 \chi \boldsymbol{E} = (n_r + in_i)^2 \boldsymbol{E},$$

where ε_0 is the electric constant, X is the (complex) electric susceptibility, and $n_r(n_i)$ is the real(imaginary) component of the refractive index of the medium. The effects of a parametric process will effect only n_r, whereas a nonzero value of $n_i$can only be caused by a non-parametric process.

Thus in linear optics a parametric process will act as a lossless dielectric with the following effects:

- Refraction
- Diffraction
- Elastic scattering
- Rayleigh scattering
- Mie scattering.

Alternatively, non-parametric processes often involve loss (or gain) and give rise to:

- Absorption
- Inelastic scattering
- Raman scattering
- Brillouin scattering
- Various optical emission processes
- Photoluminescence
- Fluorescence
- Luminescence
- Phosphorescence.

Nonlinear Optics

In a nonlinear media, the dielectric polarisation P responds nonlinearly to the electric field E of the light. As a parametric process is in general coherent, many parametric nonlinear processes will depend on phase matching and will usually bepolarisation dependent.

Sample parametric nonlinear processes:

- Second harmonic generation (SHG), or *frequency doubling*, generation of light with a doubled frequency (half the wavelength)
- Third harmonic generation (THG), generation of light with a tripled frequency (one-third the wavelength) (usually done in two steps: SHG followed by SFG of original and frequency-doubled waves)
- High harmonic generation (HHG), generation of light with frequencies much greater than the original (typically 100 to 1000 times greater)
- Sum frequency generation (SFG), generation of light with a frequency that is the sum of two other frequencies (SHG is a special case of this)
- Difference frequency generation (DFG), generation of light with a frequency that is the difference between two other frequencies
- Optical parametric amplification (OPA), amplification of a signal input in the presence of a higher-frequency pump wave, at the same time generating an *idler* wave (can be considered as DFG)
- Optical parametric oscillation (OPO), generation of a signal and idler wave using a parametric amplifier in a resonator (with no signal input)
- Optical parametric generation (OPG), like parametric oscillation but without a resonator, using a very high gain instead
- Spontaneous parametric down conversion (SPDC), the amplification of the vacuum fluctuations in the low gain regime
- Optical Kerr effect, intensity dependent refractive index
- Four-wave mixing (FWM)
- Self-focusing

- Kerr-lens modelocking (KLM)
- Self-phase modulation (SPM), a $X^{(3)}$ effect
- Optical solitons
- Cross-phase modulation (XPM)
- Four-wave mixing (FWM), can also arise from other nonlinearities
- Cross-polarised wave generation (XPW), a $X^{(3)}$ effect in which a wave with polarisation vector perpendicular to the input is generated

Sample non-parametric nonlinear processes:

- Stimulated Raman scattering
- Raman amplification
- Two-photon absorption, simultaneous absorption of two photons, transferring the energy to a single electron
- Multiphoton absorption
- Multiple photoionization, near-simultaneous removal of many bound electrons by one photon.

Optical Amplifier

An optical amplifier is a device that amplifies an optical signal directly, without the need to first convert it to an electrical signal. An optical amplifier may be thought of as a laser without an optical cavity, or one in which feedback from the cavity is suppressed. Optical amplifiers are important in optical communication and laser physics.

There are several different physical mechanisms that can be used to amplify a light signal, which correspond to the major types of optical amplifiers. In doped fibre amplifiers and bulk lasers, stimulated emission in the amplifier's gain mediumcauses amplification of incoming light. In semiconductor optical amplifiers (SOAs), electron-hole recombination occurs. In Raman amplifiers, Raman scattering of incoming light with phonons in the lattice of the gain medium produces photonscoherent with the incoming photons. Parametric amplifiers use parametric amplification.

Laser Amplifiers

Almost any laser active gain medium can be pumped to produce gain for light at the wavelength of a laser made with the same material as its gain medium. Such amplifiers are commonly used to produce high power laser systems. Special types such as regenerative amplifiers and chirped-pulse amplifiers are used to amplify ultrashort pulses.

Doped Fibre Amplifiers

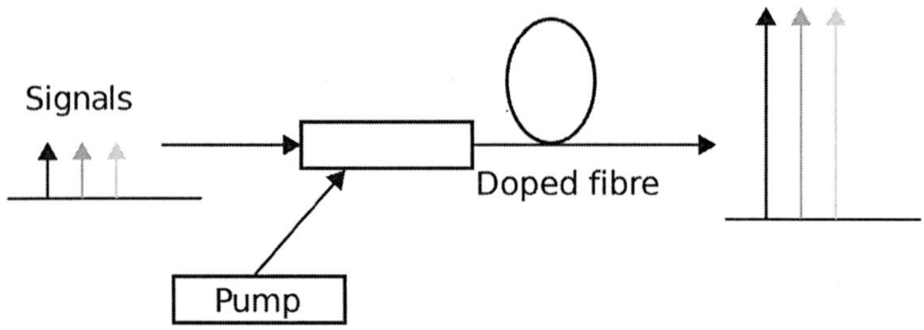

Figure 4: Schematic diagram of a simple Doped Fibre Amplifier

Doped fibre amplifiers (DFAs) are optical amplifiers that use a doped optical fibre as a gain medium to amplify an optical signal. They are related to fibre lasers. The signal to be amplified and a pump laser are multiplexed into the doped fibre, and the signal is amplified through interaction with the doping ions. The most common example is the Erbium Doped Fibre Amplifier (EDFA), where the core of a silica fibre is doped with trivalent Erbium ions and can be efficiently pumped with a laser at a wavelength of 980 nm or 1,480 nm, and exhibits gain in the 1,550 nm region.

Amplification is achieved by stimulated emission of photons from dopant ions in the doped fibre. The pump laser excites ions into a higher energy from where they can decay via stimulated emission of a photon at the signal wavelength back to a lower energy level. The excited ions can also decay spontaneously (spontaneous emission) or even through nonradiative processes involving interactions with phonons of the glass matrix. These last two decay mechanisms compete with stimulated emission reducing the efficiency of light amplification.

The *amplification window* of an optical amplifier is the range of optical wavelengths for which the amplifier yields a usable gain. The amplification window is determined by the spectroscopic properties of the dopant ions, the glass structure of the optical fibre, and the wavelength and power of the pump laser.

Although the electronic transitions of an isolated ion are very well defined, broadening of the energy levels occurs when the ions are incorporated into the glass of the optical fibre and thus the amplification window is also broadened. This broadening is both homogeneous (all ions exhibit the same broadened spectrum) and inhomogeneous (different ions in different glass locations exhibit different spectra). Homogeneous broadening arises from the interactions with phonons of the glass, while inhomogeneous broadening is caused by differences in the glass sites where different ions are hosted.

Different sites expose ions to different local electric fields, which shifts the energy levels via the Stark effect. In addition, the Stark effect also removes the degeneracy of energy states having the same total angular momentum (specified by the quantum number J). Thus, for example, the trivalent Erbium ion (Er^{+3}) has a ground state with J = 15/2, and in the presence of an electric field splits into J + 1/2 = 8 sublevels with slightly different energies.

The first excited state has J = 13/2 and therefore a Stark manifold with 7 sublevels. Transitions from the J = 13/2 excited state to the J= 15/2 ground state are responsible for the gain at 1.5 μm wavelength. The gain spectrum of the EDFA has several peaks that are smeared by the above broadening mechanisms. The net result is a very broad spectrum (30 nm in silica, typically). The broad gain-bandwidth of fibre amplifiers make them particularly useful in wavelength-division multiplexed communications systems as a single amplifier can be utilised to amplify all signals being carried on a fibre and whose wavelengths fall within the gain window.

Basic Principle of EDFA

A relatively high-powered beam of light is mixed with the input signal using a wavelength selective coupler. The input signal and the excitation light must of course be at significantly different wavelengths. The mixed light is guided into a section of fibre with erbium ions included in the core. This high-powered light beam excites the erbium ions to their higher-energy state.

When the photons belonging to the signal at a different wavelength from the pump light meet the excited erbium atoms, the erbium atoms give up some of their energy to the signal and return to their lower-energy state. A significant point is that the erbium gives up its energy in the form of additional photons which are exactly in the same phase and direction as the signal being amplified. So the signal is amplified along its direction of travel only. This is not unusual - when an atom "lases" it always gives up its energy in the same direction and phase as the incoming light. That is just the way lasers work. Thus all of the additional signal power is guided in the same fibre mode as the incoming signal. There is usually an isolator placed at the output to prevent reflections returning from the attached fibre. Such reflections disrupt amplifier operation and in the extreme case can cause the amplifier to become a laser.

Noise

The principal source of noise in DFAs is Amplified Spontaneous Emission (ASE), which has a spectrum approximately the same as the gain spectrum of the amplifier.

As well as decaying via stimulated emission, electrons in the upper energy level can also decay by spontaneous emission, which occurs at random, depending upon the glass structure and inversion level. Photons are emitted spontaneously in all directions, but a proportion of those will be emitted in a direction that falls within the numerical aperture of the fibre and are thus captured and guided by the fibre. Those photons captured may then interact with other dopant ions, and are thus amplified by stimulated emission. The initial spontaneous emission is therefore amplified in the same manner as the signals, hence the term *Amplified Spontaneous Emission*. ASE is emitted by the amplifier in both the forward and reverse directions, but only the forward ASE is a direct concern to system performance since that noise will co-propagate with the signal to the receiver where it degrades system performance. Counter-propagating ASE can, however, lead to degradation of the amplifier's performance since the ASE can deplete the inversion level and thereby reduce the gain of the amplifier.

Gain Saturation

Gain is achieved in a DFA due to population inversion of the dopant ions. The inversion level of a DFA is set, primarily, by the power of the pump wavelength and the power at the amplified wavelengths. As the signal power increases, or the pump power decreases, the inversion level will reduce and thereby the gain of the amplifier will be reduced. This effect is known as gain saturation – as the signal level increases, the amplifier saturates and cannot produce any more output power, and therefore the gain reduces. Saturation is also commonly known as gain compression.

To achieve optimum noise performance DFAs are operated under a significant amount of gain compression (10 dB typically), since that reduces the rate of spontaneous emission, thereby reducing ASE.

Another advantage of operating the DFA in the gain saturation region is that small fluctuations in the input signal power are reduced in the output amplified signal: smaller input signal powers experience larger (less saturated) gain, while larger input powers see less gain.

The leading edge of the pulse is amplified, until the saturation energy of the gain medium is reached. In some condition, the width (FWHM) of the pulse is reduced.

Inhomogeneous Broadening Effects

Due to the inhomogeneous portion of the linewidth broadening of the dopant ions, the gain spectrum has an inhomogeneous component and gain saturation occurs, to a small extent, in an inhomogeneous manner. This effect is known as *Spectral hole burning* because a high power signal at one wavelength can 'burn' a hole in the gain for wavelengths close to that signal by saturation of the inhomogeneously broadened ions. Spectral holes vary in width depending on the characteristics of the optical fibre in question and the power of the burning signal, but are typically less than 1 nm at the short wavelength end of the C-band, and a few nm at the long wavelength end of the C-band. The depth of the holes are very small, though, making it difficult to observe in practice.

Polarisation Effects

Although the DFA is essentially a polarisation independent amplifier, a small proportion of the dopant ions interact preferentially with certain polarisations and a small dependence on the polarisation of the input signal may occur (typically < 0.5 dB). This is called Polarisation Dependent Gain (PDG). The absorption and emission crossections of the ions can be modelled as ellipsoids with the major axes aligned at random in all directions in different glass sites. The random distribution of the orientation of the ellipsoids in a glass produces a macroscopically isotropic medium, but a strong pump laser induces an anisotropic distribution by selectively exciting those ions that are more aligned with the optical field vector of the pump. Also, those excited ions aligned with the signal field produce more stimulated emission. The change in gain is thus dependent on the alignment of the polarisations of the pump and signal lasers – i.e. whether the two lasers are interacting with the same subset of dopant ions or not. In an ideal doped fibre without birefringence, the PDG would be inconveniently large. Fortunately, in optical fibers small amounts of birefringence are always present and, furthermore, the fast and slow axes vary randomly along the fibre length. A typical DFA has several tens of meters, long enough to already show this randomness of the birefringence axes. These two combined effects (which in transmission fibers give rise to Polarisation Mode Dispersion) produce a misalignment of the relative polarisations of the signal and pump lasers along the fibre, thus tending to average out the PDG. The result is that PDG is very difficult to observe in a single amplifier (but is noticeable in links with several cascaded amplifiers).

Erbium-doped Fibre Amplifiers

The erbium-doped fibre amplifier (EDFA) is the most deployed fibre amplifier as its amplification window coincides with the third transmission window of silica-based optical fibre.

Two bands have developed in the third transmission window – the *Conventional*, or C-band, from approximately 1525 nm – 1565 nm, and the *Long*, or L-band, from approximately 1570 nm to 1610 nm. Both of these bands can be amplified by EDFAs, but it is normal to use two different amplifiers, each optimised for one of the bands.

The principal difference between C- and L-band amplifiers is that a longer length of doped fibre is used in L-band amplifiers. The longer length of fibre allows a lower inversion level to be used, thereby giving at longer wavelengths (due to the band-structure of Erbium in silica) while still providing a useful amount of gain.

EDFAs have two commonly-used pumping bands – 980 nm and 1480 nm. The 980 nm band has a higher absorption cross-section and is generally used where low-noise performance is required. The absorption band is relatively narrow and so wavelength stabilised laser sources are typically needed. The 1480 nm band has a lower, but broader, absorption cross-section and is generally used for higher power amplifiers. A combination of 980 nm and 1480 nm pumping is generally utilised in amplifiers.

The optical fibre amplifier was invented by H. J. Shaw and Michel Digonnet at Stanford University, California, in the early 1980s. The EDFA was first demonstrated several years later by a group including David N. Payne, R. Mears, and L. Reekie, from the University of Southampton and a group from AT&T Bell Laboratories, E. Desurvire, P. Becker, and J. Simpson.

Doped Fibre Amplifiers for other Wavelength Ranges

Thulium doped fibre amplifiers have been used in the S-band (1450–1490 nm) and Praseodymium doped amplifiers in the 1300 nm region. However, those regions have not seen any significant commercial use so far and so those amplifiers have not been the subject of as much development as the EDFA. However, Ytterbium doped fibre lasers and amplifiers, operating near 1 micrometre wavelength, have many applications in industrial processing of materials, as these devices can be made with extremely high output power (tens of kilowatts).

Semiconductor Optical Amplifier

Semiconductor optical amplifiers (SOAs) are amplifiers which use a semiconductor to provide the gain medium. These amplifiers have a similar structure to Fabry–Pérot laser diodes but with anti-reflection design elements at the endfaces. Recent designs include anti-reflective coatings and tilted waveguide and window regions which can reduce endface reflection to less than 0.001%. Since this creates a loss of power from the cavity which is greater than the gain it prevents the amplifier from acting as a laser.

Semiconductor optical amplifiers are typically made from group III-V compound semiconductors such as GaAs/AlGaAs, InP/InGaAs, InP/InGaAsP and InP/InAlGaAs, though any direct band gap semiconductors such as II-VI could conceivably be used. Such amplifiers are often used in telecommunication systems in the form of fibre-pigtailed components, operating at signal wavelengths between 0.85 μm and 1.6 μm and generating gains of up to 30 dB.

The semiconductor optical amplifier is of small size and electrically pumped. It can be potentially less expensive than the EDFA and can be integrated with semiconductor lasers, modulators, etc. However, the performance is still not comparable with the EDFA. The SOA has higher noise, lower gain, moderate polarisation dependence and high nonlinearity with fast transient time. The main advantage of SOA is that all four types of nonlinear operations (cross gain modulation, cross phase modulation, wavelength conversion and four wave mixing) can be conducted. Furthermore, SOA can be run with a low power laser. This originates from the short nanosecond or less upper state lifetime, so that the gain reacts rapidly to changes of pump or signal power and the changes of gain also cause phase changes which can distort the signals. This nonlinearity presents the most severe problem for optical communication applications. However it provides the possibility for gain in different wavelength regions from the EDFA. "Linear optical amplifiers" using gain-clamping techniques have been developed.

High optical nonlinearity makes semiconductor amplifiers attractive for all optical signal processing like all-optical switching and wavelength conversion. There has been much research on semiconductor optical amplifiers as elements for optical signal processing, wavelength conversion, clock recovery, signal demultiplexing, and pattern recognition.

Vertical-cavity SOA

A recent addition to the SOA family is the vertical-cavity SOA (VCSOA). These devices are similar in structure to, and share many features with, vertical-cavity surface-emitting lasers (VCSELs). The major difference when comparing VCSOAs and VCSELs is the reduced mirror reflectivities used in the amplifier cavity.

With VCSOAs, reduced feedback is necessary to prevent the device from reaching lasing threshold. Due to the extremely short cavity length, and correspondingly thin gain medium, these devices exhibit very low single-pass gain (typically on the order of a few percent) and also a very large free spectral range (FSR). The small single-pass gain requires relatively high mirror reflectivities to boost the total signal gain. In addition to boosting the total signal gain, the use of the resonant cavity structure results in a very narrow gain bandwidth; coupled with the large FSR of the optical cavity, this effectively limits operation of the VCSOA to single-channel amplification. Thus, VCSOAs can be seen as amplifying filters.

Given their vertical-cavity geometry, VCSOAs are resonant cavity optical amplifiers that operate with the input/output signal entering/exiting normal to the wafer surface. In addition to their small size, the surface normal operation of VCSOAs leads to a number of advantages, including low power consumption, low noise figure, polarisation insensitive gain, and the ability to fabricate high fill factor two-dimensional arrays on a single semiconductor chip.

These devices are still in the early stages of research, though promising preamplifier results have been demonstrated. Further extensions to VCSOA technology are the demonstration of wavelength tunable devices.

These MEMS-tunable vertical-cavity SOAs utilise a microelectromechanical systems (MEMS) based tuning mechanism for wide and continuous tuning of the peak gain wavelength of the amplifier. SOAs has a more rapid gain response ,which is in the order of 1 to 100ps.

Raman Amplifier

In a Raman amplifier, the signal is intensified by Raman amplification. Unlike the EDFA and SOA the amplification effect is achieved by a nonlinear interaction between the signal and a pump laser within an optical fibre.

There are two types of Raman amplifier: distributed and lumped. A distributed Raman amplifier is one in which the transmission fibre is utilised as the gain medium by multiplexing a pump wavelength with signal wavelength, while a lumped Raman amplifier utilises a dedicated, shorter length of fibre to provide amplification. In the case of a lumped Raman amplifier highly nonlinear fibre with a small core is utilised to increase the interaction between signal and pump wavelengths and thereby reduce the length of fibre required.

The pump light may be coupled into the transmission fibre in the same direction as the signal (co-directional pumping), in the opposite direction (contra-directional pumping) or both. Contra-directional pumping is more common as the transfer of noise from the pump to the signal is reduced.

The pump power required for Raman amplification is higher than that required by the EDFA, with in excess of 500 mW being required to achieve useful levels of gain in a distributed amplifier. Lumped amplifiers, where the pump light can be safely contained to avoid safety implications of high optical powers, may use over 1W of optical power.

The principal advantage of Raman amplification is its ability to provide distributed amplification within the transmission fibre, thereby increasing the length of spans between amplifier and regeneration sites. The amplification bandwidth of Raman amplifiers is defined by the pump wavelengths utilised and so amplification can be provided over wider, and different, regions than may be possible with other amplifier types which rely on dopants and device design to define the amplification 'window'.

Note: The text of an earlier version of this article was taken from the public domain Federal Standard 1037C.

Optical Parametric Amplifier

An optical parametric amplifier allows the amplification of a weak Signal-Impulse in a noncentrosymmetric nonlinear medium (e.g. BBO). In contrast to the previously mentioned amplifiers, which are mostly used in telecommunication environments, this type finds its main application in expanding the frequency tunability of ultrafast solid-state lasers (e.g. Ti:sapphire). By using a noncollinear interaction geometry Optical Parametric Amplifiers are capable of extreme broad amplification bandwidths.

Eckart Conditions

The Eckart conditions, named after Carl Eckart, sometimes referred to as Sayvetz conditions, simplify the nuclear motion (rovibrational) Schrödinger equation that arises in the second step of the Born-Oppenheimer approximation. The Eckart conditions allow to a large extent the separation of the external (rotation and translation) motions from the internal (vibration) motions. Although the rotational and vibrational motions of the nuclei in a molecule cannot be fully separated, the Eckart conditions minimise the coupling between these two.

Definition of Eckart Conditions

The Eckart conditions can only be formulated for a semi-rigid molecule, which is a molecule with a potential energy surface $V(R_1, R_2,..R_N)$ that has a well-defined minimum for R_A^0 ($A = 1,\ldots,N$). These equilibrium coordinates of the nuclei—with masses M_A—are expressed with respect to a fixed orthonormal principal axes frame and hence satisfy the relations

$$\sum_{A=1}^{N} M_A\left(\delta_{ij} \mid \boldsymbol{R}_A^0 \mid^2 - R_{Ai}^0 R_{Aj}^0\right) = \lambda_i^0 \delta_{ij} \quad \text{and} \quad \sum_{A=1}^{N} M_A \boldsymbol{R}_A^0 = \boldsymbol{0}.$$

Here λ_i^0 is a principal inertia moment of the equilibrium molecule. The triplets $R_A^0 = (R_{A1}^0, R_{A2}^0, R_{A3}^0)$ satisfying these conditions, enter the theory as a given set of real constants. Following Biedenharn and Louck we introduce an orthonormal body-fixed frame, the *Eckart frame*,

$$\vec{\boldsymbol{F}} = \{\vec{f}_1, \vec{f}_2, \vec{f}_3\}.$$

If we were tied to the Eckart frame, which—following the molecule—rotates and translates in space, we would observe the molecule in its equilibrium geometry when we would draw the nuclei at the points,

$$\vec{R}_A^0 \equiv \vec{\boldsymbol{F}}\cdot\boldsymbol{R}_A^0 = \sum_{i=1}^{3} \vec{f}_i R_{Ai}^0, \quad A = 1,\ldots,N$$

Let the elements of R_A be the coordinates with respect to the Eckart frame of the position vector of nucleus A ($A = 1,\ldots,N$). Since we take the origin of the Eckart frame in the instantaneous center of mass, the following relation

$$\sum_A M_A \boldsymbol{R}_A = \boldsymbol{0}$$

holds. We define *displacement coordinates*

$$\boldsymbol{d}_A \equiv \boldsymbol{R}_A - \boldsymbol{R}_A^0 .$$

Clearly the displacement coordinates satisfy the translational Eckart conditions,

$$\sum_{A=1}^{N} M_A \boldsymbol{d}_A = 0.$$

The rotational Eckart conditions for the displacements are:

$$\sum_{A=1}^{N} M_A \boldsymbol{R}_A^0 \times \boldsymbol{d}_A = 0,$$

where indicates a vector product. These rotational conditions follow from the specific construction of the Eckart frame.

Finally, for a better understanding of the Eckart frame it may be useful to remark that it becomes a principal axes frame in the case that the molecule is a rigid rotor, that is, when all N displacement vectors are zero.

Separation of External and Internal Coordinates

The N *position vectors* $\vec{R}_A$ *of the nuclei constitute a* 3N *dimensional linear space* R^{3N}: *the* configuration space. *The Eckart conditions give an orthogonal direct sum decomposition of this space*

$$\boldsymbol{R}^{3N} = \boldsymbol{R}_{\text{ext}} \oplus \boldsymbol{R}_{\text{int}} .$$

The elements of the $3N$-6 dimensional subspace R_{int} are referred to as *internal coordinates*, because they are invariant under overall translation and rotation of the molecule and, thus, depend only on the internal (vibrational) motions. The elements of the 6-dimensional subspace R_{ext} are referred to as *external coordinates*, because they are associated with the overall translation and rotation of the molecule.

To clarify this nomenclature we define first a basis for R_{ext}. To that end we introduce the following 6 vectors (i=1,2,3):

$$\vec{s}_i^{\,A} \equiv \vec{f}_i$$

$$\vec{s}_{i+3}^{\,A} \equiv \vec{f}_i \times \vec{R}_A^0 .$$

An orthogonal, unnormalised, basis for R_{ext} is,

$$\vec{S}_t \equiv \text{row}(\sqrt{M_1}\, \vec{s}_t^{\,1},\ldots,\sqrt{M_N}\, \vec{s}_t^{\,N}) \quad \text{for} \quad t=1,\ldots,6.$$

A mass-weighted displacement vector can be written as

$$\vec{D} \equiv \text{col}(\sqrt{M_1}\, \vec{d}^{\,1},\ldots,\sqrt{M_N}\, \vec{d}^{\,N}) \quad \text{with} \quad \vec{d}^{\,A} \equiv \vec{\boldsymbol{F}}\cdot\boldsymbol{d}_A.$$

For i=1,2,3,

$$\vec{S}_i\cdot\vec{D} = \sum_{A=1}^{N} M_A \vec{s}_i^{\,A}\cdot\vec{d}^{\,A} = \sum_{A=1}^{N} M_A d_{Ai} = 0,$$

where the zero follows because of the translational Eckart conditions. For i=4,5,6

$$\vec{S}_i\cdot\vec{D} = \sum_{A=1}^{N} M_A\left(\vec{f}_i\times\vec{R}_A^{0}\right)\cdot\vec{d}^{\,A} = \vec{f}_i\cdot\sum_{A=1}^{N} M_A \vec{R}_A^{0}\times\vec{d}^{\,A} = \sum_{A=1}^{N} M_A\left(\boldsymbol{R}_A^{0}\times\boldsymbol{d}_A\right.$$

where the zero follows because of the rotational Eckart conditions. We conclude that the displacement vector $\vec{D}$ belongs to the orthogonal complement of R_{ext}, so that it is an internal vector.

We obtain a basis for the internal space by defining 3N-6 linearly independent vectors

$$\vec{Q}_r \equiv \text{row}(\frac{1}{\sqrt{M_1}}\,\vec{q}_r^{\,1},\ldots,\frac{1}{\sqrt{M_N}}\,\vec{q}_r^{\,N}), \quad \text{for} \quad r=1,\ldots,3N-6.$$

The vectors $\vec{q}_r^{\,A}$ could be Wilson's s-vectors or could be obtained in the harmonic approximation by diagonalizing the Hessian of V. We next introduce internal (vibrational) modes,

$$q_r \equiv \vec{Q}_r\cdot\vec{D} = \sum_{A=1}^{N} \vec{q}_r^{\,A}\cdot\vec{d}^{\,A} \quad \text{for} \quad r=1,\ldots,3N-6.$$

The physical meaning of q_r depends on the vectors . For instance, q_r could be a symmetric stretching mode, in which two C—H bonds are simultaneously stretched and contracted. We already saw that the corresponding external modes are zero because of the Eckart conditions,

$$s_t \equiv \vec{S}_t\cdot\vec{D} = \sum_{A=1}^{N} M_A\, \vec{s}_t^{\,A}\cdot\vec{d}^{\,A} = 0 \quad \text{for} \quad t=1,\ldots,6.$$

Overall Translation and Rotation

The vibrational (internal) modes are invariant under translation and infinitesimal rotation of the equilibrium (reference) molecule if and only if the Eckart conditions apply. This will be shown in this subsection.

An overall translation of the reference molecule is given by

$$\vec{R}_A^0 \mapsto \vec{R}_A^0 + \vec{t}\,,$$

for any arbitrary 3-vector $\vec{t}$. An infinitesimal rotation of the molecule is given by

$$\vec{R}_A^0 \mapsto \vec{R}_A^0 + \Delta\varphi\,(\vec{n}\times\vec{R}_A^0)$$

where Δφ is an infinitesimal angle, $\Delta\varphi \gg (\Delta\varphi)^2$, and is an arbitrary unit vector. From the orthogonality of to the external space follows that the satisfy

$$\sum_{A=1}^{N}\vec{q}_r^{\,A} = \vec{0} \quad \text{and} \quad \sum_{A=1}^{N}\vec{R}_A^0\times\vec{q}_r^{\,A} = \vec{0}.$$

Clearly, $\vec{q}_r^{\,A}$ is invariant under translation if and only if

$$\sum_A \vec{q}_r^{\,A} = 0,$$

because the vector $\vec{t}$ is arbitrary. So, the translational Eckart conditions imply the translational invariance of the vectors belonging to internal space and conversely. Under rotation we have,

$$q_r \mapsto \sum_A \vec{q}_r^{\,A}\cdot\left(\vec{d}^A - \Delta\varphi\,(\vec{n}\times\vec{R}_A^0)\right) = q_r - \Delta\varphi\,\vec{n}\cdot\sum_A \vec{R}_A^0\times\vec{q}_r^{\,A} = q$$

Rotational invariance follows if and only if

$$\sum_A \vec{R}_A^0\times\vec{q}_r^{\,A}.$$

The external modes, on the other hand, are *not* invariant and it is not difficult to show that they change under translation as follows:

$$s_i \mapsto s_i + M\vec{f}_i\cdot\vec{t} \quad \text{for} \quad i = 1,2,3$$

$$s_i \mapsto s_i \quad \text{for} \quad i = 4,5,6,$$

where M is the total mass of the molecule. They change under infinitesimal rotation as follows

$$s_i \mapsto s_i \quad \text{for} \quad i = 1,2,3$$

$$s_i \mapsto s_i + \Delta\phi\vec{f}_i\cdot\boldsymbol{I}^0\cdot\vec{n} \quad \text{for} \quad i = 4,5,6,$$

where I^0 is the inertia tensor of the equilibrium molecule. This behavior shows that the first three external modes describe the overall translation of the molecule, while the modes 4, 5, and, 6 describe the overall rotation.

Vibrational Energy

The vibrational energy of the molecule can be written in terms of coordinates with respect to the Eckart frame as

$$2T_{\text{vib}} = \sum_{A=1}^{N} M_A \dot{\boldsymbol{R}}_A \cdot \dot{\boldsymbol{R}}_A = \sum_{A=1}^{N} M_A \dot{\boldsymbol{d}}_A \cdot \dot{\boldsymbol{d}}_A .$$

Because the Eckart frame is non-inertial, the total kinetic energy comprises also centrifugal and Coriolis energies. These stay out of the present discussion. The vibrational energy is written in terms of the displacement coordinates, which are linearly dependent because they are contaminated by the 6 external modes, which are zero, i.e., the d_A's satisfy 6 linear relations. It is possible to write the vibrational energy solely in terms of the internal modes q_r (r =1, ..., 3N-6) as we will now show. We write the different modes in terms of the displacements

$$q_r = \sum_{Aj} d_{Aj} \quad \left(q_{rj}^{A}\right)$$

$$s_i = \sum_{Aj} d_{Aj} \quad \left(M_A \delta_{ij}\right) = 0$$

$$s_{i+3} = \sum_{Aj} d_{Aj} \quad \left(M_A \sum_{k} \epsilon_{ikj} R_{Ak}^{0}\right) = 0$$

The parenthesized expressions define a matrix B relating the internal and external modes to the displacements. The matrix B may be partitioned in an internal (3N-6 x 3N) and an external (6 x 3N) part,

$$\boldsymbol{v} \equiv \begin{pmatrix} q_1 \\ \vdots \\ \vdots \\ q_{3N-6} \\ 0 \\ \vdots \\ 0 \end{pmatrix} = \begin{pmatrix} \boldsymbol{B}^{\text{int}} \\ \cdots \\ \boldsymbol{B}^{\text{ext}} \end{pmatrix} \boldsymbol{d} \equiv \boldsymbol{Bd}.$$

We define the matrix M by

$$\boldsymbol{M} \equiv \text{diag}(\boldsymbol{M}_1, \boldsymbol{M}_2, \ldots, \boldsymbol{M}_N) \quad \text{and} \quad \boldsymbol{M}_A \equiv \text{diag}(M_A, M_A, M_A$$

and from the relations given in the previous sections follow the matrix relations

$$B^{ext}M^{-1}(B^{ext})^{T} = \text{diag}(N_1,\ldots,N_6) \equiv N,$$

and

$$B^{int}M^{-1}(B^{ext})^{T} = 0.$$

We define

$$G \equiv B^{int}M^{-1}(B^{int})^{T}.$$

By using the rules for block matrix multiplication we can show that

$$(B^{T})^{-1}MB^{-1} = \begin{pmatrix} G^{-1} & 0 \\ 0 & N^{-1} \end{pmatrix},$$

where G^{-1} is of dimension (3*N*-6 x 3*N*-6) and N^{-1} is (6 x 6). The kinetic energy becomes

$$2T_{vib} = \dot{d}^{T} M\dot{d} = \dot{v}^{T} (B^{T})^{-1}MB^{-1} \dot{v} = \sum_{r,r'=1}^{3N-6} (G^{-1})_{rr'}\dot{q}_r\dot{q}_{r'}$$

where we used that the last 6 components of v are zero. This form of the kinetic energy of vibration enters Wilson's GF method. It is of some interest to point out that the potential energy in the harmonic approximation can be written as follows

$$2V_{harm} = d^{T}Hd = v^{T}(B^{T})^{-1}HB^{-1}v = \sum_{r,r'=1}^{3N-6} F_{rr'}q_r q_{r'},$$

where H is the Hessian of the potential in the minimum and F, defined by this equation, is the F matrix of the GF method.

Relation to the Harmonic Approximation

In the harmonic approximation to the nuclear vibrational problem, expressed in displacement coordinates, one must solve the generalised eigenvalue problem

$$HC = MC\Phi,$$

where H is a $3N \times 3N$ symmetric matrix of second derivatives of the potential $V(R_1,R_2,\ldots,R_N)$. H is the Hessian matrix of *V* in the equilibrium $R_1^0,\ldots,R_N^0$. The diagonal matrix M contains the masses on the diagonal. The diagonal matrix Φ contains the eigenvalues, while the columns of C contain the eigenvectors.

It can be shown that the invariance of V under simultaneous translation over t of all nuclei implies that vectors T = (t, ... , t) are in the kernel of H. From the invariance of V under an infinitesimal rotation of all nuclei around s, it can be shown that also the vectors S = (s x R_1^0, ..., s x R_N^0) are in the kernel of H :

$$\boldsymbol{H}\begin{pmatrix} \boldsymbol{t} \\ \vdots \\ \boldsymbol{t} \end{pmatrix} = \begin{pmatrix} \boldsymbol{0} \\ \vdots \\ \boldsymbol{0} \end{pmatrix} \quad \text{and} \quad \boldsymbol{H}\begin{pmatrix} \boldsymbol{s}\times\boldsymbol{R}_1^0 \\ \vdots \\ \boldsymbol{s}\times\boldsymbol{R}_N^0 \end{pmatrix} = \begin{pmatrix} \boldsymbol{0} \\ \vdots \\ \boldsymbol{0} \end{pmatrix}$$

Thus, six columns of C corresponding to eigenvalue zero are determined algebraically. (If the generalised eigenvalue problem is solved numerically, one will find in general six linearly independent linear combinations of S and T). The eigenspace corresponding to eigenvalue zero is at least of dimension 6 (often it is exactly of dimension 6, since the other eigenvalues, which are force constants, are never zero for molecules in their ground state). Thus, T and S correspond to the overall (external) motions: translation and rotation, respectively. They are *zero-energy modes* because space is homogeneous (force-free) and isotropic (torque-free).

By the definition in this article, the non-zero frequency modes are internal modes, since they are within the orthogonal complement of R_{ext}. The generalised orthogonalities: $\boldsymbol{C}^{T}\boldsymbol{M}\boldsymbol{C} = \boldsymbol{I}$ applied to the "internal" (non-zero eigenvalue) and "external" (zero-eigenvalue) columns of C are equivalent to the Eckart conditions.

GF Method

The GF method, sometimes referred to as FG method, is a classical mechanical method introduced by E. Bright Wilson to obtain certain *internal coordinates* for a vibrating semi-rigid molecule, the so-called *normal coordinates* Q_k. Normal coordinates decouple the classical vibrational motions of the molecule and thus give an easy route to obtaining vibrational amplitudes of the atoms as a function of time. In Wilson's GF method it is assumed that the molecular kinetic energy consists only of harmonic vibrations of the atoms, *i.e.,* overall rotational and translational energy is ignored. Normal coordinates appear also in a quantum mechanical description of the vibrational motions of the molecule and the Coriolis coupling between rotations and vibrations.

It follows from application of the Eckart conditions that the matrix G<sp>-1 gives the kinetic energy in terms of arbitrary linear internal coordinates, while F represents the (harmonic) potential energy in terms of these coordinates. The GF method gives the linear transformation from general internal coordinates to the special set of normal coordinates.

The GF Method

A non-linear molecule consisting of N atoms has 3N-6 internal degrees of freedom, because positioning a molecule in three-dimensional space requires three degrees of freedom and the description of its orientation in space requires another three degree of freedom. These degrees of freedom must be subtracted from the 3N degrees of freedom of a system of N particles.

The atoms in a molecule are bound by a potential energy surface (PES) (or a force field) which is a function of 3N-6 coordinates. The internal degrees of freedom q1, ..., q3N-6 describing the PES in an optimum way are often non-linear; they are for instance valence coordinates, such as bending and torsion angles and bond stretches. It is possible to write the quantum mechanical kinetic energy operator for such curvilinear coordinates, but it is hard to formulate a general theory applicable to any molecule. This is why Wilson linearised the internal coordinates by assuming small displacements. The linearised version of the internal coordinate qt is denoted by St.

The PES V can be Taylor expanded around its minimum in terms of the S_t. The third term (the Hessian of V) evaluated in the minimum is a force derivative matrix F. In the harmonic approximation the Taylor series is ended after this term. The second term, containing first derivatives, is zero because it is evaluated in the minimum of V. The first term can be included in the zero of energy. Thus,

$$2V \approx \sum_{s,t=1}^{3N-6} F_{st} S_s S_t \, .$$

The classical vibrational kinetic energy has the form:

$$2T = \sum_{s,t=1}^{3N-6} g_{st}(\boldsymbol{q}) \dot{S}_s \dot{S}_t ,$$

where g_{st} is an element of the metric tensor of the internal (curvilinear) coordinates. The dots indicate time derivatives. Mixed terms $S_s \dot{S}_t$ generally present in curvilinear coordinates are not present here, because only linear coordinate transformations are used. Evaluation of the metric tensor g in the minimum q^0 of V gives the positive definite and symmetric matrix G = $g(q^0)^{-1}$. One can solve the following two matrix problems simultaneously

$$\boldsymbol{L}^{\mathrm{T}}\boldsymbol{FL} = \boldsymbol{\Phi} \quad \text{and} \quad \boldsymbol{L}^{\mathrm{T}}\boldsymbol{G}^{-1}\boldsymbol{L} = \boldsymbol{E},$$

since they are equivalent to the generalised eigenvalue problem

$$\boldsymbol{GFL} = \boldsymbol{L\Phi},$$

where $\boldsymbol{\Phi} = \mathrm{diag}(f_1,\ldots,f_{3N-6})$ where f_i is equal to $4\pi^2 v_i^2$ (υ_i is the frequency of normal mode *i*); $\boldsymbol{E}$ is the unit matrix. The matrix L^{-1} contains the *normal coordinates* Q_k in its rows:

$$Q_k = \sum_{t=1}^{3N-6} (\boldsymbol{L}^{-1})_{kt} S_t, \quad k = 1,\ldots,3N-6.$$

Because of the form of the generalised eigenvalue problem, the method is called the GF method, often with the name of its originator attached to it: Wilson's GF method. By matrix transposition in both sides of the equation and using the fact that both G and F are symmetric matrices, as are diagonal matrices, one can recast this equation into a very similar one for FG . This is why the method is also referred to as Wilson's FG method.

We introduce the vectors

$$\boldsymbol{s} = \mathrm{col}(S_1,\ldots,S_{3N-6}) \quad \text{and} \quad \boldsymbol{Q} = \mathrm{col}(Q_1,\ldots,Q_{3N-6}),$$

which satisfy the relation

$$\boldsymbol{s} = \boldsymbol{LQ}.$$

Upon use of the results of the generalised eigenvalue equation, the energy $E = T + V$ (in the harmonic approximation) of the molecule becomes,

$$2E = \dot{\boldsymbol{s}}^{T}\boldsymbol{G}^{-1}\dot{\boldsymbol{s}} + \boldsymbol{s}^{\mathrm{T}}\boldsymbol{Fs}$$

$$= \dot{\boldsymbol{Q}}^{T}\left(\boldsymbol{L}^{\mathrm{T}}\boldsymbol{G}^{-1}\boldsymbol{L}\right)\dot{\boldsymbol{Q}} + \boldsymbol{Q}^{\mathrm{T}}\left(\boldsymbol{L}^{\mathrm{T}}\boldsymbol{FL}\right)\boldsymbol{Q}$$

$$= \dot{\boldsymbol{Q}}^{T}\dot{\boldsymbol{Q}} + \boldsymbol{Q}^{\mathrm{T}} \rightarrow \boldsymbol{Q} = \sum_{t=1}^{3N-6}\left(\dot{Q}_t^2 + f_t Q_t^2\right).$$

The Lagrangian $L = T - V$ is

$$L = \frac{1}{2}\sum_{t=1}^{3N-6}\left(\dot{Q}_t^2 - f_t Q_t^2\right).$$

The corresponding Lagrange equations are identical to the Newton equations

$$\ddot{Q}_t + f_t Q_t = 0$$

for a set of uncoupled harmonic oscillators. These ordinary second-order differential equations are easily solved, yielding Q_t as a function of time; see the article on harmonic oscillators.

Normal Coordinates in Terms of Cartesian Displacement Coordinates

Often the normal coordinates are expressed as linear combinations of Cartesian displacement coordinates. Let R_A be the position vector of nucleus A and R_A^0 the corresponding equilibrium position.

Then $\boldsymbol{d}_A \equiv \boldsymbol{R}_A - \boldsymbol{R}_A^0$ is by definition the Cartesian displacement coordinate of nucleus A. Wilson's linearising of the internal curvilinear coordinates q_t expresses the coordinate S_t in terms of the displacement coordinates

$$S_t = \sum_{A=1}^{N}\sum_{i=1}^{3} s_{Ai}^t d_{Ai} = \sum_{A=1}^{N} \boldsymbol{s}_A^t \cdot \boldsymbol{d}_A, \quad \text{for} \quad t = 1,\ldots,3N-6,$$

where s_A^t is known as a *Wilson s-vector*. If we put the s_{Ai}^t into a $3N$-6 x $3N$ matrix B, this equation becomes in matrix language

$$\boldsymbol{s} = \boldsymbol{Bd}.$$

The actual form of the matrix elements of B can be fairly complicated. Especially for a torsion angle, which involves 4 atoms, it requires tedious vector algebra to derive the corresponding values of the .

Known as the *Wilson s-vector method*, the book by Wilson *et al.*, or molecular vibration. Now,

$$\boldsymbol{Q} = \boldsymbol{L}^{-1}\boldsymbol{s} = \boldsymbol{L}^{-1}\boldsymbol{Bd} \equiv \boldsymbol{Dd}.$$

In summation language:

$$Q_k = \sum_{A=1}^{N}\sum_{i=1}^{3} D_{Ai}^k d_{Ai} \quad \text{for} \quad k = 1,\ldots,3N-6.$$

Here D is a $3N$-6 x $3N$ matrix which is given by (i) the linearisation of the internal coordinates q (an algebraic process) and (ii) solution of Wilson's GF equations (a numeric process).

Relation with Eckart Conditions

From the invariance of the internal coordinates S_t under overall rotation and translation of the molecule, follows the same for the linearised coordinates s^t_A. It can be shown that this implies that the following 6 conditions are satisfied by the internal coordinates,

$$\sum_{A=1}^{N} \boldsymbol{s}^t_A = 0 \quad \text{and} \quad \sum_{A=1}^{N} \boldsymbol{R}^0_A \times \boldsymbol{s}^t_A = 0, \quad t = 1,\ldots,3N-6.$$

These conditions follow from the Eckart conditions that hold for the displacement vectors,

$$\sum_{A=1}^{N} M_A\, \boldsymbol{d}_A = 0 \quad \text{and} \quad \sum_{A=1}^{N} M_A\, \boldsymbol{R}^0_A \times \boldsymbol{d}_A = 0.$$

Chapter 6

Gas Chromatography–mass Spectrometry

Gas chromatography–mass spectrometry (GC-MS) is a method that combines the features of gas-liquid chromatography and mass spectrometry to identify different substances within a test sample. Applications of GC-MS include drug detection, fire investigation, environmental analysis, explosives investigation, and identification of unknown samples. GC-MS can also be used in airport security to detect substances in luggage or on human beings. Additionally, it can identify trace elements in materials that were previously thought to have disintegrated beyond identification.

GC-MS has been widely heralded as a "gold standard" for forensic substance identification because it is used to perform a *specific test*. A specific test positively identifies the actual presence of a particular substance in a given sample. A *non-specific test* merely indicates that a substance falls into a category of substances. Although a non-specific test could statistically suggest the identity of the substance, this could lead to false positive identification.

History

The use of a mass spectrometer as the detector in gas chromatography was developed during the 1950s by Roland Gohlke and Fred McLafferty. These sensitive devices were bulky, fragile, and originally limited to laboratory settings.

The development of affordable and miniaturised computers has helped in the simplification of the use of this instrument, as well as allowed great improvements in the amount of time it takes to analyse a sample. In 1964, Electronic Associates, Inc. (EAI), a leading U.S. supplier of analog computers, began development of a computer controlled quadruple mass spectrometer under the direction of Robert

E. Finnigan. By1966 Finnigan and collaborator Mike Uthe's EAI division had sold over 500 quadrupole residual gas-analyser instruments. In 1967, the Finnigan Instrument Corporation was formed and in early 1968, delivered the first prototype quadrapole GC/MS instruments to Stanford and Purdue University. FIC was eventually renamed Finnigan Corporation and went on to establish itself as the worldwide leader in GC/MS systems.

In 1996 the top-of-the-line high-speed GC-MS units completed analysis of fire accelerants in less than 90 seconds, whereas first-generation GC-MS would have required at least 16 minutes. By the 2000s computerised GC/MS instruments using quadrupole technology had become both essential to chemical research and one of the foremost instruments used for organic analysis. Today computerised GC/MS instruments are widely used in environmental monitoring of water, air, and soil; in the regulation of agriculture and food safety; and in the discovery and production of medicine.

Instrumentation

Figure 1: The insides of the GC-MS, with the column of the gas chromatograph in the oven on the right.

The GC-MS is composed of two major building blocks: the gas chromatograph and the mass spectrometer. The gas chromatograph utilises a capillary column which depends on the column's dimensions (length, diameter, film thickness) as well as the phase properties (e.g. 5% phenyl polysiloxane). The difference in the chemical properties between differentmolecules in a mixture will separate the molecules as the sample travels the length of the column. The molecules are retained by the column and then elute (come off of) from the column at different times (called the retention time), and this allows the mass spectrometer downstream to capture, ionize, accelerate, deflect, and detect the ionized molecules separately. The mass spectrometer does this by breaking each molecule into ionized fragments and detecting these fragments using their mass to charge ratio.

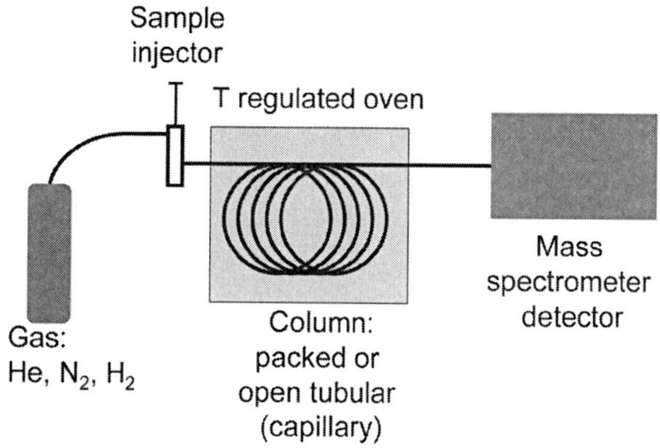

Figure 2: GC-MS schematic

These two components, used together, allow a much finer degree of substance identification than either unit used separately. It is not possible to make an accurate identification of a particular molecule by gas chromatography or mass spectrometry alone. The mass spectrometry process normally requires a very pure sample while gas chromatography using a traditional detector (e.g. Flame Ionization Detector) detects multiple molecules that happen to take the same amount of time to travel through the column (*i.e.* have the same retention time) which results in two or more molecules to co-elute. Sometimes two different molecules can also have a similar pattern of ionized fragments in a mass spectrometer (mass spectrum). Combining the two processes reduces the possibility of error, as it is extremely unlikely that two different molecules will behave in the same way in both a gas chromatograph and a mass spectrometer. Therefore, when an identifying mass spectrum appears at a characteristic retention time in a GC-MS analysis, it typically lends to increased certainty that the analyte of interest is in the sample.

Purge and Trap GC-MS

For the analysis of volatile compounds a Purge and Trap (P&T) concentrator system may be used to introduce samples. The target analytes are extracted and mixed with water and introduced into an airtight chamber. An inert gas such as Nitrogen (N_2) is bubbled through the water; this is known as purging. The volatile compounds move into the headspace above the water and are drawn along a pressure gradient (caused by the introduction of the purge gas) out of the chamber. The volatile compounds are drawn along a heated line onto a 'trap'. The trap is a column of adsorbent material at ambient temperature that holds the compounds by returning them to the liquid phase. The trap is then heated and the sample compounds are introduced to the GC-MS column via a volatiles interface, which is a split inlet system. P&T GC-MS is particularly suited to volatile organic compounds (VOCs) and BTEX compounds (aromatic compounds associated with petroleum).

Types of Mass Spectrometer Detectors

The most common type of mass spectrometer (MS) associated with a gas chromatograph (GC) is the quadrupole mass spectrometer, sometimes referred to by the Hewlett-Packard (now Agilent) trade name "Mass Selective Detector" (MSD). Another relatively common detector is the ion trap mass spectrometer. Additionally one may find a magnetic sector mass spectrometer, however these particular instruments are expensive and bulky and not typically found in high-throughput service laboratories. Other detectors may be encountered such as time of flight (TOF), tandem quadrupoles (MS-MS), or in the case of an ion trap MS^n where n indicates the number mass spectrometry stages.

Analysis

A mass spectrometer is typically utilised in one of two ways: Full Scan or Selective Ion Monitoring (SIM). The typical GC-MS instrument is capable of performing both functions either individually or concomitantly, depending on the setup of the particular instrument.

Full Scan MS

When collecting data in the full scan mode, a target range of mass fragments is determined and put into the instrument's method. An example of a typical broad range of mass fragments to monitor would be m/z 50 to m/z 400. The determination of what range to use is largely dictated by what one anticipates being in the sample while being cognizant of the solvent and other possible interferences. A MS should not be set to look for mass fragments too low or else one may detect air (found as m/z 28 due to nitrogen), carbon dioxide (m/z 44) or other possible interferences. Additionally if one is to use a large scan range then sensitivity of the instrument is decreased due to performing fewer scans per second since each scan will have to detect a wide range of mass fragments.

Full scan is useful in determining unknown compounds in a sample. It provides more information than SIM when it comes to confirming or resolving compounds in a sample. During instrument method development it may be common to first analyse test solutions in full scan mode to determine the retention time and the mass fragment fingerprint before moving to a SIM instrument method.

Selected ion Monitoring

In selected ion monitoring (SIM) certain ion fragments are entered into the instrument method and only those mass fragments are detected by the mass spectrometer. The advantages of SIM are that the detection limit is lower since the instrument is only looking at a small number of fragments (e.g. three fragments) during each scan. More scans can take place each second. Since only a few mass fragments of interest are being monitored, matrix interferences are typically lower. To additionally confirm the likelihood of a potentially positive result, it is relatively important to be sure that the ion ratios of the various mass fragments are comparable to a known reference standard.

Types of Ionization

After the molecules travel the length of the column, pass through the transfer line and enter into the mass spectrometer they are ionized by various methods with typically only one method being used at any given time. Once the sample is fragmented it will then be detected, usually by an electron multiplier diode, which essentially turns the ionized mass fragment into an electrical signal that is then detected.

The ionization technique chosen is independent of using Full Scan or SIM.

Electron Ionization

By far the most common and perhaps standard form of ionization is electron ionization (EI). The molecules enter into the MS (the source is a quadrupole or the ion trap itself in an ion trap MS) where they are bombarded with free electrons emitted from a filament, not much unlike the filament one would find in a standard light bulb. The electrons bombard the molecules, causing the molecule to fragment in a characteristic and reproducible way. This "hard ionization" technique results in the creation of more fragments of low mass to charge ratio (m/z) and few, if any, molecules approaching the molecular mass unit. Hard ionization is considered by mass spectrometrists as the employ of molecular electron bombardment, whereas "soft ionization" is charge by molecular collision with an introduced gas. The molecular fragmentation pattern is dependant upon the electron energy applied to the system, typically 70 eV (electron Volts). The use of 70 eV facilitates comparison of generated spectra with library spectra using manufacturer-supplied software or software developed by the National Institute of Standards (NIST-USA). Spectral library searches employ matching algorithms such as Probability Based Matching and dot-product matching that are used with methods of analysis written by many method standardisation agencies. Sources of libraries include NIST, Wiley, the AAFS, and instrument manufacturers. Library coverage can be checked at Compound Search.

Chemical Ionization

In chemical ionization a reagent gas, typically methane or ammonia is introduced into the mass spectrometer. Depending on the technique (positive CI or negative CI) chosen, this reagent gas will interact with the electrons and analyte and cause a 'soft' ionization of the molecule of interest. A softer ionization fragments the molecule to a lower degree than the hard ionization of EI. One of the main benefits of using chemical ionization is that a mass fragment closely corresponding to the molecular weight of the analyte of interest is produced.

Positive Chemical Ionization

In Positive Chemical Ionization (PCI) the reagent gas interacts with the target molecule, most often with a proton exchange. This produces the species in relatively high amounts.

Negative Chemical Ionization

In Negative Chemical Ionization (NCI) the reagent gas decreases the impact of the free electrons on the target analyte. This decreased energy typically leaves the fragment in great supply.

The primary goal of instrument analysis is to quantify an amount of substance. This is done by comparing the relative concentrations among the atomic masses in the generated spectrum. Two kinds of analysis are possible, comparative and original. Comparative analysis essentially compares the given spectrum to a spectrum library to see if its characteristics are present for some sample in the library. This is best performed by a computer because there are a myriad of visual distortions that can take place due to variations in scale. Computers can also simultaneously correlate more data (such as the retention times identified by GC), to more accurately relate certain data.

Another method of analysis measures the peaks in relation to one another. In this method, the tallest peak is assigned 100% of the value, and the other peaks being assigned proportionate values. All values above 3% are assigned. The total mass of the unknown compound is normally indicated by the parent peak. The value of this parent peak can be used to fit with a chemical formula containing the various elements which are believed to be in the compound. The isotope pattern in the spectrum, which is unique for elements that have many isotopes, can also be used to identify the various elements present. Once a chemical formula has been matched to the spectrum, the molecular structure and bonding can be identified, and must be consistent with the characteristics recorded by GC-MS. Typically, this identification done automatically by programs which come with the instrument, given a list of the elements which could be present in the sample.

A "full spectrum" analysis considers all the "peaks" within a spectrum. Conversely, selective ion monitoring (SIM) only monitors selected peaks associated with a specific substance. This is done on the assumption that at a given retention time, a set of ions is characteristic of a certain compound. This is a fast and efficient analysis, especially if the analyst has previous information about a sample or is only looking for a few specific substances. When the amount of information collected about the ions in a given gas chromatographic peak decreases, the sensitivity of the analysis increases. So, SIM analysis allows for a smaller quantity of a compound to be detected and measured, but the degree of certainty about the identity of that compound is reduced.

GC-tandem MS

When a second phase of mass fragmentation is added, for example using a second quadrupole in a quadrupole instrument, it is called tandem MS (MS/MS). MS/MS can sometimes be used to quantitate low levels of target compounds in the presence of a high sample matrix background.

The first quadrupole (Q1) is connected with a collision cell (q2) and another quadrupole (Q3). Both quadrupoles can be used in scanning or static mode, depending on the type of MS/MS analysis being performed. Types of analysis include product ion scan, precursor ion scan, Selected Reaction Monitoring (SRM) (sometimes referred to as Multiple Reaction Monitoring (MRM)) and Neutral Loss Scan. For example: When Q1 is in static mode (looking at one mass only as in

SIM), and Q3 is in scanning mode, one obtains a so-called product ion spectrum (also called "daughter spectrum"). From this spectrum, one can select a prominent product ion which can be the product ion for the chosen precursor ion. The pair is called a "transition" and forms the basis for SRM. SRM is highly specific and virtually eliminates matrix background.

Applications

Environmental Monitoring and Cleanup

GC-MS is becoming the tool of choice for tracking organic pollutants in the environment. The cost of GC-MS equipment has decreased significantly, and the reliability has increased at the same time, which has contributed to its increased adoption in environmental studies. There are some compounds for which GC-MS is not sufficiently sensitive, including certain pesticides and herbicides, but for most organic analysis of environmental samples, including many major classes of pesticides, it is very sensitive and effective.

Criminal Forensics

GC-MS can analyse the particles from a human body in order to help link a criminal to a crime. The analysis of fire debris using GC-MS is well established, and there is even an established American Society for Testing Materials (ASTM) standard for fire debris analysis. GCMS/MS is especially useful here as samples often contain very complex matrices and results, used in court, need to be highly accurate.

Law Enforcement

GC-MS is increasingly used for detection of illegal narcotics, and may eventually supplant drug-sniffing dogs. It is also commonly used in forensic toxicology to find drugs and/or poisons in biological specimens of suspects, victims, or the deceased.

Sports Anti-doping Analysis

GC-MS is the main tool used in sports anti-doping laboratories to test athletes' urine samples for prohibited performance-enhancing drugs, for example anabolic steroids.

Security

A post–September 11 development, explosive detection systems have become a part of all US airports. These systems run on a host of technologies, many of them based on GC-MS. There are only three manufacturers certified by the FAA to provide these systems, one of which is Thermo Detection (formerly Thermedics), which produces the EGIS, a GC-MS-based line of explosives detectors. The other two manufacturers are Barringer Technologies, now owned by Smith's Detection Systems, and Ion Track Instruments, part of General Electric Infrastructure Security Systems.

Food, Beverage and Perfume Analysis

Foods and beverages contain numerous aromatic compounds, some naturally present in the raw materials and some forming during processing. GC-MS is extensively used for the analysis of these compounds which include esters, fatty acids, alcohols, aldehydes, terpenes etc. It is also used to detect and measure contaminants from spoilage or adulteration which may be harmful and which is often controlled by governmental agencies, for example pesticides.

Astrochemistry

Several GC-MS have left earth. Two were brought to Mars by the Viking program. Venera 11 and 12 and Pioneer Venus analysed the atmosphere of Venus with GC-MS. The Huygens probe of the Cassini-Huygens mission landed one GC-MS on Saturn's largest moon, Titan. The material in the comet 67P/Churyumov-Gerasimenko will be analysed by the Rosetta mission with a chiral GC-MS in 2014.

Medicine

Dozens of congenital metabolic diseases also known as Inborn error of metabolism are now detectable by newborn screening tests, especially the testing using gas chromatography–mass spectrometry. GC-MS can determine compounds in urine even in minor concentration. These compounds are normally not present but appear in individuals suffering with metabolic disorders. This is an increasingly becoming a common way to diagnose IEM for earlier diagnosis and institution of treatment eventually leading to a better outcome. It is now possible to test a newborn for over 100 genetic metabolic disorders by a urine test at birth based on GC-MS.

In combination with isotopic labelling of metabolic compounds, the GC-MS is used for determining metabolic activity. Most applications are based on the use of ^{13}C as the labelling and the measurement of ^{13}C-^{12}C ratios with an isotope ratio mass spectrometer (IRMS); an MS with a detector designed to measure a few select ions and return values as ratios.

High-performance Liquid Chromatography

High-performance liquid chromatography (sometimes referred to as high-pressure liquid chromatography), HPLC, is a chromatographic technique that can separate a mixture of compounds and is used in biochemistry and analytical chemistry to identify, quantify and purify the individual components of the mixture.

HPLC typically utilises different types of stationary phases, a pump that moves the mobile phase(s) and analyte through the column, and a detector to provide a characteristic retention time for the analyte. The detector may also provide additional information related to the analyte, (i.e. UV/Vis spectroscopic data for analyte if so equipped). Analyte retention time varies depending on the strength of its interactions with the stationary phase, the ratio/composition of solvent(s) used, and the flow rate of the mobile phase. It is a form of liquid chromatography that utilises smaller column size, smaller media inside the column, and higher mobile phase pressures.

With HPLC, a pump (rather than gravity) provides the higher pressure required to move the mobile phase and analyte through the densely packed column. The increased density arises from smaller particle sizes. This allows for a better separation on columns of shorter length when compared to ordinary column chromatography.

Operation

The sample to be analysed or separated is introduced, in small volumes, into the stream of mobile phase. The solution moved through the column is slowed by specific chemical or physical interactions with the stationary phase present within the column. The velocity of the solution depends on the nature of the sample and on the compositions of the stationary (column) phase. The time at which a specific sample elutes (comes out of the end of the column) is called the retention time; the retention time under particular conditions is considered an identifying characteristic of a given sample. The use of smaller particle size column packing (which creates higher backpressure) increases the linear velocity giving the components less time to diffuse within the column, improving the chromatogram resolution. Common solvents used include any miscible combination of water or various organic liquids (the most common are methanol and acetonitrile). Water may contain buffers or salts to assist in the separation of the sample components, or compounds such as trifluoroacetic acid which acts as an ion pairing agent.

A further refinement of HPLC is to vary the mobile phase composition during the analysis; gradient elution. A normal gradient for reversed phase chromatography might start at 5% methanol and progress linearly to 50% methanol over 25 minutes; the gradient depends on how hydrophobic the sample is. The gradient separates the sample mixtures as a function of the affinity. This partitioning process is similar to that which occurs during a liquid-liquid extraction but is continuous, not step-wise. In this example, using a water/methanol gradient, more hydrophobic components will elute (come off the column) when the mobile phase consists mostly of methanol (giving a relatively hydrophobic mobile phase).

The choice of solvents, additives and gradient depend on the nature of the column and sample. Often a series of tests are performed on the sample together with a number of trial runs in order to find the HPLC method which gives the best peak separation.

Partition Chromatography

Figure 3: A modern self contained HPLC.

Partition chromatography was the first kind of chromatography that chemists developed. The partition coefficient principle has been applied in paper chromatography, thin layer chromatography, gas phase and liquid-liquid applications. The 1952 Nobel Prize in chemistry was earned by Archer John Porter Martin and Richard Laurence Millington Synge for their development of the technique, which was used for their separation of amino acids. Partition chromatography uses a retained solvent, on the surface or within the grains or fibres of an "inert" solid supporting matrix as with paper chromatography; or takes advantage of some coulombic and/or hydrogen donor interaction with the solid support. Molecules equilibrate (partition) between a liquid stationary phase and the eluent.

Known as Hydrophilic Interaction Chromatography (HILIC) in HPLC, this method separates analytes based on polar differences. HILIC most often uses a bonded polar stationary phase and a non-polar, water miscible, mobile phase. Partition HPLC has been used historically on unbonded silica or alumina supports. Each works effectively for separating analytes by relative polar differences, however, HILIC has the advantage of separating acidic, basic and neutral solutes in a single chromatogram.

The polar analytes diffuse into a stationary water layer associated with the polar stationary phase and are thus retained. Retention strengths increase with increased analyte polarity, and the interaction between the polar analyte and the polar stationary phase (relative to the mobile phase) increases the elution time. The interaction strength depends on the functional groups in the analyte molecule which promote partitioning but can also include coulombic (electrostatic) interaction and hydrogen donor capability.

Use of more polar solvents in the mobile phase will decrease the retention time of the analytes, whereas more hydrophobic solvents tend to increase retention times.

Normal-phase Chromatosgraphy

Also known as normal-phase HPLC (NP-HPLC), or adsorption chromatography, this method separates analytes based on adsorption to a stationary surface chemistry and by polarity. It was one of the first kinds of HPLC that chemists developed. NP-HPLC uses a polar stationary phase and a non-polar, non-aqueous mobile phase, and works effectively for separating analytes readily soluble in non-polar solvents. The analyte associates with and is retained by the polar stationary phase. Adsorption strengths increase with increased analyte polarity, and the interaction between the polar analyte and the polar stationary phase (relative to the mobile phase) increases the elution time. The interaction strength depends not only on the functional groups in the analyte molecule, but also on steric factors. The effect of sterics on interaction strength allows this method to resolve (separate) structural isomers.

The use of more polar solvents in the mobile phase will decrease the retention time of the analytes, whereas more hydrophobic solvents tend to increase retention times. Very polar solvents in a mixture tend to deactivate the stationary phase by creating a stationary bound water layer on the stationary phase surface. This behavior is somewhat peculiar to normal phase because it is most purely an adsorptive mechanism (the interactions are with a hard surface rather than a soft layer on a surface).

Partition and NP-HPLC fell out of favour in the 1970s with the development of reversed-phase HPLC because of a lack of reproducibility of retention times as water or protic organic solvents changed the hydration state of the silica or aluminachromatographic media. Recently it has become useful again with the development of HILIC bonded phases which improve reproducibility.

Displacement Chromatography

The basic principle of displacement chromatography is: A molecule with a high affinity for the chromatography matrix (the displacer) will compete effectively for binding sites, and thus displace all molecules with lesser affinities. There are distinct differences between displacement and elution chromatography. In elution mode, substances typically emerge from a column in narrow, Gaussian peaks. Wide separation of peaks, preferably to baseline, is desired in order to achieve maximum purification. The speed at which any component of a mixture travels down the column in elution mode depends on many factors. But for two substances to travel at different speeds, and thereby be resolved, there must be substantial differences in some interaction between the biomolecules and the chromatography matrix. Operating parameters are adjusted to maximise the effect of this difference. In many cases, baseline separation of the peaks can be achieved only with gradient elution and low column loadings. Thus, two drawbacks to elution mode chromatography, especially at the preparative scale, are operational complexity, due to gradient solvent pumping, and low throughput, due to low column loadings. Displacement chromatography has advantages over elution chromatography in that components are resolved into consecutive zones of pure substances rather than “peaks”. Because the process takes advantage of the nonlinearity of the isotherms, a larger column feed can be separated on a given column with the purified components recovered at significantly higher concentrations.

Reversed-phase Chromatography (RPC)

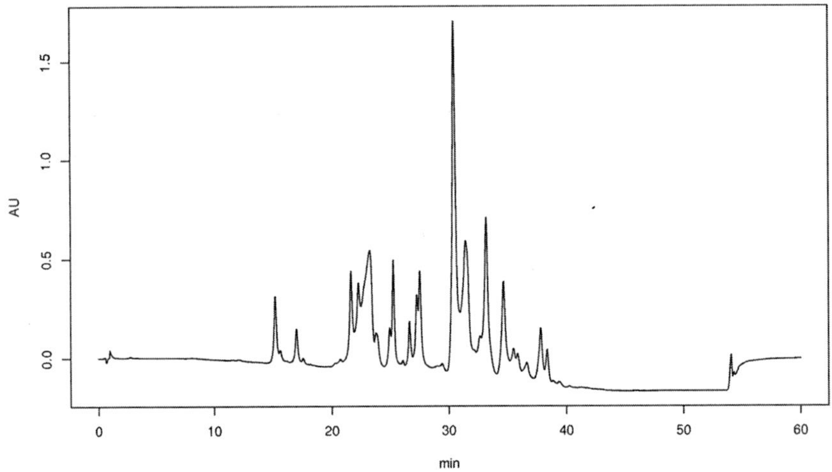

Figure 4: A chromatogram of complex mixture (perfume water) obtained by reversed phase HPLC

Reversed phase HPLC (RP-HPLC or RPC) has a non-polar stationary phase and an aqueous, moderately polar mobile phase. One common stationary phase is a silica which has been treated with RMe_2SiCl, where R is a straight chain alkyl group such as $C_{18}H_{37}$ or C_8H_{17}. With these stationary phases, retention time is longer for molecules which are less polar, while polar molecules elute more readily. An investigator can increase retention time by adding more water to the mobile phase; thereby making the affinity of the hydrophobic analyte for the hydrophobic stationary phase stronger relative to the now more hydrophilic mobile phase. Similarly, an investigator can decrease retention time by adding more organic solvent to the eluent. RPC is so commonly used that it is often incorrectly referred to as "HPLC" without further specification. The pharmaceutical industry regularly employs RPC to qualify drugs before their release.

RPC operates on the principle of hydrophobic forces, which originate from the high symmetry in the dipolar water structure and play the most important role in all processes in life science. RPC allows the measurement of these interactive forces. The binding of the analyte to the stationary phase is proportional to the contact surface area around the non-polar segment of the analyte molecule upon association with the ligand in the aqueous eluent. This solvophobic effect is dominated by the force of water for "cavity-reduction" around the analyte and the C_{18}-chain versus the complex of both. The energy released in this process is proportional to the surface tension of the eluent (water: 7.3×10^{-6} J/cm^2, methanol: 2.2×10^{-6} J/cm^2) and to the hydrophobic surface of the analyte and the ligand respectively. The retention can be decreased by adding a less polar solvent (methanol, acetonitrile) into the mobile phase to reduce the surface tension of water. Gradient elution uses this effect by automatically reducing the polarity and the surface tension of the aqueous mobile phase during the course of the analysis.

Structural properties of the analyte molecule play an important role in its retention characteristics. In general, an analyte with a larger hydrophobic surface area (C-H, C-C, and generally non-polar atomic bonds, such as S-S and others) results in a longer retention time because it increases the molecule's non-polar surface area, which is non-interacting with the water structure. On the other hand, polar groups, such as -OH, $-NH_2$, COO^- or $-NH_3^+$ reduce retention as they are well integrated into water. Very large molecules, however, can result in an incomplete interaction between the large analyte surface and the ligand's alkyl chains and can have problems entering the pores of the stationary phase.

Retention time increases with hydrophobic (non-polar) surface area. Branched chain compounds elute more rapidly than their corresponding linear isomers because the overall surface area is decreased. Similarly organic compounds with single C-C-bonds elute later than those with a C=C or C-C-triple bond, as the double or triple bond is shorter than a single C-C-bond.

Aside from mobile phase surface tension (organisational strength in eluent structure), other mobile phase modifiers can affect analyte retention. For example, the addition of inorganic salts causes a moderate linear increase in the surface tension of aqueous solutions (ca. 1.5×10^{-7} J/cm^2 per Mol for NaCl, 2.5×10^{-7} J/cm^2 per Mol for $(NH_4)_2SO_4$), and because the entropy of the analyte-solvent interface is controlled by surface tension, the addition of salts tend to increase the retention time. This technique is used for mild separation and recovery of proteins and protection of their biological activity in protein analysis (hydrophobic interaction chromatography, HIC). Another important component is the influence of the pH since this can change the hydrophobicity of the analyte. For this reason most methods use a buffering agent, such as sodium phosphate, to control the pH. The buffers serve multiple purposes: they control pH, neutralise the charge on any residual exposed silica on the stationary phase and act as ion pairing agents to neutralise charge on the analyte. Ammonium formate is commonly added in mass spectrometry to improve detection of certain analytes by the formation of ammonium adducts.

A volatile organic acid such as acetic acid, or most commonly formic acid, is often added to the mobile phase if mass spectrometry is used to analyse the column effluent. Trifluoroacetic acid is used infrequently in mass spectrometry applications due to its persistence in the detector and solvent delivery system, but can be effective in improving retention of analytes such as carboxylic acids in applications utilising other detectors, as it is one of the strongest organic acids. The effects of acids and buffers vary by application but generally improve the chromatography.

Reversed phase columns are quite difficult to damage compared with normal silica columns; however, many reversed phase columns consist of alkyl derivatized silica particles and should never be used with aqueous bases as these will destroy the underlying silica particle. They can be used with aqueous acid, but the column should not be exposed to the acid for too long, as it can corrode the metal parts of the HPLC equipment. RP-HPLC columns should be flushed with clean solvent after use to remove residual acids or buffers, and stored in an appropriate composition of solvent. The metal content of HPLC columns must be kept low if the best possible ability to separate substances is to be retained. A good test for the metal content of a column is to inject a sample which is a mixture of 2,2'- and 4,4'-bipyridine. Because the 2,2'-bipy can chelate the metal, the shape of the peak for the 2,2'-bipy will be distorted (tailed) when metal ions are present on the surface of the silica...

Size-exclusion Chromatography

Size-exclusion chromatography (SEC), also known as *gel permeation chromatography* or *gel filtration chromatography*, separates particles on the basis of size. It is generally a low resolution chromatography and thus it is often reserved for the final, "polishing" step of a purification. It is also useful for determining the tertiary structure and quaternary structure of purified proteins. SEC is used primarily for the analysis of large molecules such as proteins or polymers. SEC works by trapping these smaller molecules in the pores of a particle. The larger molecules simply pass by the pores as they are too large to enter the pores. Larger molecules therefore flow through the column quicker than smaller molecules, that is, the smaller the molecule, the longer the retention time.

This technique is widely used for the molecular weight determination of polysaccharides. SEC is the official technique (suggested by European pharmacopeia) for the molecular weight comparison of different commercially available low-molecular weight heparins.

Ion-exchange Chromatography

In ion-exchange chromatography (IC), retention is based on the attraction between solute ions and charged sites bound to the stationary phase. Ions of the same charge are excluded. Types of ion exchangers include:

- Polystyrene resins – These allow cross linkage which increases the stability of the chain. Higher cross linkage reduces swerving, which increases the equilibration time and ultimately improves selectivity.
- Cellulose and dextran ion exchangers (gels) – These possess larger pore sizes and low charge densities making them suitable for protein separation.
- Controlled-pore glass or porous silica. In general, ion exchangers favour the binding of ions of higher charge and smaller radius.

An increase in counter ion (with respect to the functional groups in resins) concentration reduces the retention time. An increase in pH reduces the retention time in cation exchange while a decrease in pH reduces the retention time in anion exchange. This form of chromatography is widely used in the following applications: water purification, preconcentration of trace components, ligand-exchange chromatography, ion-exchange chromatography of proteins, high-pH anion-exchange chromatography of carbohydrates and oligosaccharides, and others.

Bioaffinity Chromatography

This chromatographic process relies on the property of biologically active substances to form stable, specific, and reversible complexes. The formation of these complexes involves the participation of common molecular forces such as theVan der Waals interaction, electrostatic interaction, dipole-dipole interaction, hydrophobic interaction, and the hydrogen bond. An efficient, biospecific bond is formed by a simultaneous and concerted action of several of these forces in the complementary binding sites.

Aqueous Normal-phase Chromatography

Aqueous normal-phase chromatography (ANP) is a chromatographic technique which encompasses the mobile phase region between reversed-phase chromatography (RP) and organic normal phase chromatography (ONP). This technique is used to achieve unique selectivity for hydrophilic compounds, showing normal phase elution using reverse-phase solvents.

Isocratic Flow and Gradient Elution

A separation in which the mobile phase composition remains constant throughout the procedure is termed isocratic (meaning constant composition). The word was coined by Csaba Horvath who was one of the pioneers of HPLC.,

The mobile phase composition does not have to remain constant. A separation in which the mobile phase composition is changed during the separation process is described as a gradient elution. One example is a gradient starting at 10% methanol and ending at 90% methanol after 20 minutes. The two components of the mobile phase are typically termed "A" and "B"; A is the "weak" solvent which allows the solute to elute only slowly, while B is the "strong" solvent which rapidly elutes the solutes from the column. In reverse-phase chromatography, solvent A is often water or an aqueous buffer, while B is an organic solvent miscible with water, such as acetonitrile, methanol, THF, or isopropanol.

In isocratic elution, peak width increases with retention time linearly according to the equation for N, the number of theoretical plates. This leads to the disadvantage that late-eluting peaks get very flat and broad. Their shape and width may keep them from being recognised as peaks.

Gradient elution decreases the retention of the later-eluting components so that they elute faster, giving narrower (and taller) peaks for most components. This also improves the peak shape for tailed peaks, as the increasing concentration of the organic eluent pushes the tailing part of a peak forward. This also increases the peak height (the peak looks "sharper"), which is important in trace analysis.

The gradient program may include sudden "step" increases in the percentage of the organic component, or different slopes at different times – all according to the desire for optimum separation in minimum time.

In isocratic elution, the selectivity does not change if the column dimensions (length and inner diameter) change – that is, the peaks elute in the same order. In gradient elution, the elution order may change as the dimensions or flow rate change.

The driving force in reversed phase chromatography originates in the high order of the water structure. The role of the organic component of the mobile phase is to reduce this high order and thus reduce the retarding strength of the aqueous component.

Parameters

Internal Diameter: The internal diameter (ID) of an HPLC column is an important parameter that influences the detection sensitivity and separation selectivity in gradient elution. It also determines the quantity of analyte that can be loaded onto the column. Larger columns are usually seen in industrial applications, such as the purification of a drug product for later use. Low-ID columns have improved sensitivity and lower solvent consumption at the expense of loading capacity.

- Larger ID columns (over 10 mm) are used to purify usable amounts of material because of their large loading capacity.
- Analytical scale columns (4.6 mm) have been the most common type of columns, though smaller columns are rapidly gaining in popularity. They are used in traditional quantitative analysis of samples and often use a UV-Vis absorbance detector.
- Narrow-bore columns (1–2 mm) are used for applications when more sensitivity is desired either with special UV-vis detectors, fluorescence detection or with other detection methods like liquid chromatography-mass spectrometry
- Capillary columns (under 0.3 mm) are used almost exclusively with alternative detection means such as mass spectrometry. They are usually made from fused silica capillaries, rather than the stainless steel tubing that larger columns employ.

Particle Size

Most traditional HPLC is performed with the stationary phase attached to the outside of small spherical silica particles (very small beads). These particles come in a variety of sizes with 5 ìm beads being the most common. Smaller particles generally provide more surface area and better separations, but the pressure required for optimum linear velocity increases by the inverse of the particle diameter squared.

This means that changing to particles that are half as big, keeping the size of the column the same, will double the performance, but increase the required pressure by a factor of four. Larger particles are used in preparative HPLC (column diameters 5 cm up to >30 cm) and for non-HPLC applications such as solid-phase extraction.

Pore Size

Many stationary phases are porous to provide greater surface area. Small pores provide greater surface area while larger pore size has better kinetics, especially for larger analytes. For example, a protein which is only slightly smaller than a pore might enter the pore but does not easily leave once inside.

Pump Pressure

Pumps vary in pressure capacity, but their performance is measured on their ability to yield a consistent and reproducible flow rate. Pressure may reach as high as 40 MPa (6000 lbf/in^2), or about 400 atmospheres. Modern HPLC systems have been improved to work at much higher pressures, and therefore are able to use much smaller particle sizes in the columns (<2 ìm). These "Ultra High Performance Liquid Chromatography" systems or RSLC/UHPLCs can work at up to 100 MPa (15,000 lbf/in^2), or about 1000 atmospheres. The term "UPLC" is a trademark of the Waters Corporation, but is sometimes used to refer to the more general technique.

Refractive Index

In optics the refractive index or index of refraction of a substance or medium is a measure of the speed of light in that medium. It is expressed as a ratio of the speed of light in vacuum relative to that in the considered medium. This can be written mathematically as:

n = speed of light in a vacuum / speed of light in medium.

For example, the refractive index of water is 1.33, meaning that light travels 1.33 times as fast in vacuum as it does in water.

As light moves from a medium, such as air, water, or glass, into another it may change its propagation direction in proportion to the change in refractive index. This refraction is governed by Snell's law, and is illustrated in the figure to the right. Refractive index of materials varies with the wavelength of light. This is called dispersion and results in a slightly different refractive index for eachcolor.

The wavelength λ of light in a material is determined by the refractive index according to $\lambda = \lambda_0 / n$, where λ_0 is the wavelength of the light in vacuum. Brewster's angle, the critical angle for total internal reflection, and the reflectivity of a surface is also affected by the refractive index. These material parameters can be calculated using the Fresnel equations.

The concept of refractive index is widely used within the full electromagnetic spectrum, from x-rays to radio waves. It can also be used with wave phenomena other than light, e.g. sound. In this case the speed of sound is used instead of that of light and a reference medium other than vacuum must be chosen.

Definitions

The refractive index, n, of a medium is defined as the ratio of the speed, c, of a wave phenomenon such as light or sound in a reference medium to the phase speed, v_p, of the wave in the medium in question:

$$n = \frac{c}{v_{\mathrm{p}}}.$$

It is most commonly used in the context of light with vacuum as a reference medium, although historically other reference media (e.g. air at a standardised pressure and temperature) have been common. It is usually given the symbol n. In the case of light, it equals

$$n = \sqrt{\grave{o}_r . \mu_r},$$

where is the material's relative permittivity, and $\grave{\imath}_r$ is its relative permeability. For most naturally occurring materials, $\grave{\imath}_r$ is very close to 1 at optical frequencies, therefore n is approximately . Contrary to a widespread misconception, the real part of a complex n may be less than one, depending upon the material and wavelength. This has practical technical applications, such as effective mirrors for X-rays based on total external reflection.

The phase speed is defined as the rate at which the crests of the waveform propagate; that is, the rate at which the phase of the waveform is moving. The *group speed* is the rate at which the *envelope* of the waveform is propagating; that is, the rate of variation of the amplitude of the waveform. Provided the waveform is not distorted significantly during propagation, it is the group speed that represents the rate at which information (and energy) may be transmitted by the wave (for example, the speed at which a pulse of light travels down an optical fibre). For the analytic properties constraining the unequal phase and group speeds in dispersive media, refer to dispersion (optics).

Another common definition of the refractive index comes from the refraction of a light ray entering a medium. The refractive index is the ratio of the sines of the angles of incidence θ_1 and refraction θ_2 as light passes into the medium or mathematically

$$n = \frac{\sin \theta_1}{\sin \theta_2}.$$

The angles are measured to the normal of the surface. This definition is based on Snell's law and is equivalent to the definition above if the light enters from the reference medium (normally vacuum).

A complex refractive index is often used to take absorption into account. This is further discussed in the Dispersion and absorption section below.

A closely related quantity is *refractivity*, which in atmospheric applications is denoted N and defined as $N = 10^6(n - 1)$. The 10^6 factor is used because for air, n deviates from unity at most a few parts per thousand.

Typical Values

Selected refractive indices atλ = 589 nm (more extensive list here).

Material	*n*
Gases at 0 °C and 1 atm	
Air	1.000293
Helium	1.000036
Hydrogen	1.000132
Carbon dioxide	1.00045
Liquids at 20 °C	
Water	1.333
Ethanol	1.36
Methanol	1.329
Benzene	1.501
Glycerine	1.473
Solids	

Ice	1.309
Fused silica	1.46
PMMA	1.49
Crown glass (typical)	1.52
Flint glass (typical)	1.62
Diamond	2.42

For visible light most transparent media have refractive indices between 1 and 2. A few examples are given in the table to the right. These values are measured at the yellow doublet sodium D-line, with a wavelength of 589 nanometres, as is conventionally done. Gases at atmospheric pressure have refractive indices close to 1 because of their low density. Most plastics have refractive indices in the range from 1.3 to 1.7, but some high-refractive-index polymers can have a value as high as 1.76.

For infrared light refractive indices can be considerably higher. Germanium is transparent in this region and has a refractive index of about 4, making it an important material for infrared optics.

Refractive Index Below

A widespread misconception is that since, according to the theory of relativity, nothing can travel faster than the speed of light in vacuum, the refractive index cannot be lower than 1. This is erroneous since the refractive index measures the phase velocity of light, which does not carry energy or information, the two things limited in propagation speed. The phase velocity is the speed at which the crests of the wave move and can be faster than the speed of light in vacuum, and thereby give a refractive index below 1. This can occur close to resonance frequencies, in plasmas, and for x-rays. In the x-ray regime the refractive indices are lower than but very close to 1 (exceptions close to some resonance frequencies). As an example, water has a refractive index of 1"2.6×10"7 at a photon energy of 30 keV (0.04 nm wavelength).

Negative Refractive Index

Recent research has also demonstrated the existence of the negative refractive index, which can occur if permittivity and permeability have simultaneous negative values. This can be achieved with periodically constructed metamaterials. The resulting negative refraction (i.e., a reversal of Snell's law) offers the possibility of the superlens and other exotic phenomena.

Microscopic Explanation

At the microscale, an electromagnetic wave's phase speed is slowed in a material because the electric field creates a disturbance in the charges of each atom (primarily the electrons) proportional to the electric susceptibility of the medium. (Similarly, the magnetic field creates a disturbance proportional to the magnetic susceptibility.) As the electromagnetic fields oscillate in the wave, the charges in the material will be "shaken" back and forth at the same frequency. The charges thus radiate their own electromagnetic wave that is at the same frequency, but usually with a phase delay, as the charges may move out of phase with the force driving them. The light wave travelling in the medium is the macroscopic superposition (sum) of all such contributions in the material: The original wave plus the waves radiated by all the moving charges. This wave is typically a wave with the same frequency but shorter wavelength than the original, leading to a slowing of the wave's phase speed. Most of the radiation from oscillating material charges will modify the incoming wave, changing its velocity. However, some net energy will be radiated in other directions or even at other frequencies.

Depending on the relative phase of the original driving wave and the waves radiated by the charge motion, there are several possibilities:

- If the electrons emit a light wave which is 90° out of phase with the light wave shaking them, it will cause the total light wave to travel more slowly. This is the normal refraction of transparent materials like glass or water, and corresponds to a refractive index which is real and greater than 1.
- If the electrons emit a light wave which is 270° out of phase with the light wave shaking them, it will cause the total light wave to travel more quickly. This is called "anomalous refraction", and is observed close to absorption lines, with X-rays, and in some microwave systems. It corresponds to a refractive index less than 1. (Even though the phase velocity of light is greater than the speed of light in vacuum c, the signal velocity is not, as discussed above). If the response is sufficiently strong and out-of-phase, the result is negative refractive index discussed below.
- If the electrons emit a light wave which is 180° out of phase with the light wave shaking them, it will destructively interfere with the original light to reduce the total light intensity. This

is light absorption in opaque materials and corresponds to an imaginary refractive index.

- If the electrons emit a light wave which is in phase with the light wave shaking them, it will amplify the light wave. This is rare, but occurs in lasers due to stimulated emission. It corresponds to an imaginary index of refraction, with the opposite sign as absorption.

For most materials at visible-light frequencies, the phase is somewhere between 90° and 180°, corresponding to a combination of both refraction and absorption.

Dispersion

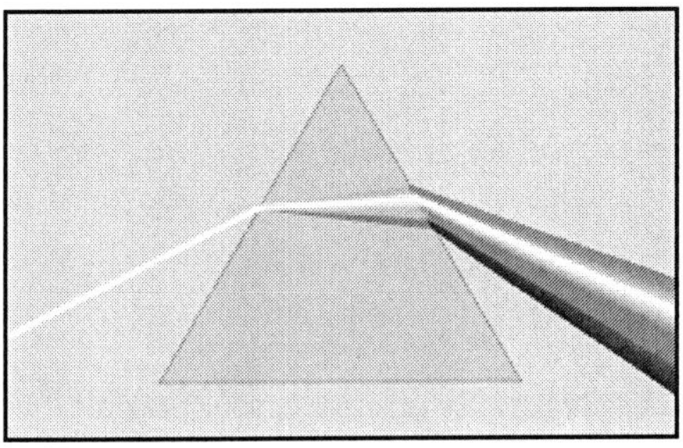

Figure 5: In a prism dispersion causes different colours to refract at different angles, splitting white light into a rainbow of colours.

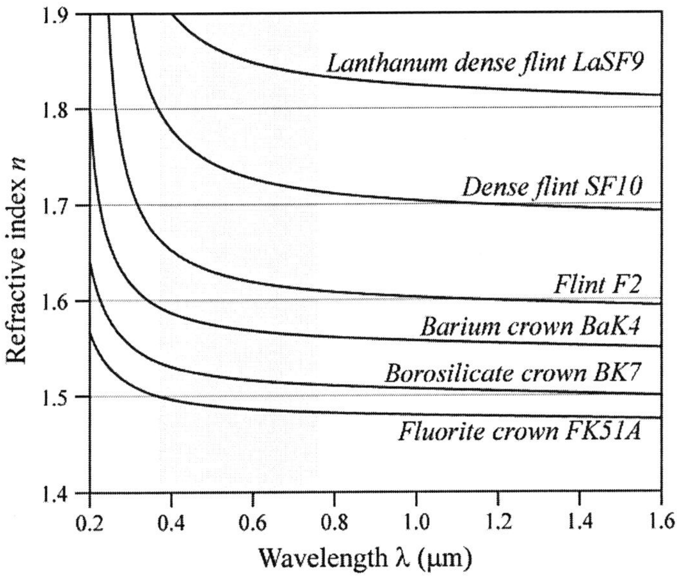

Figure 6: The variation of refractive index with wavelength for various glasses.

The refractive index of materials varies with the wavelength (and frequency) of light. This is called dispersion and causes prisms to divide white light into its constituent spectral colours, and explains how rainbows are formed. As the refractive index varies with wavelength, according to Snell's law, so will the refraction angle as light goes from one material to another. This makes different colours go in different directions. Dispersion also causes the focal length of lenses to be wavelength dependent. This is a type of chromatic aberration, which often needs to be corrected for in imaging systems.

In regions of the spectrum where the material does not absorb, the refractive index tends to decrease with increasing wavelength, and thus increase with frequency. This is called normal dispersion, in contrast to anomalous dispersion, where the refractive index increases with wavelength. For visible light normal dispersion means that the refractive index is higher for blue light than for red.

For optics in the visual range the amount of dispersion of a lens material is often quantified by the Abbe number $V = \frac{n_{\text{yellow}} - 1}{n_{\text{blue}} - n_{\text{red}}}$. For a more accurate description of the wavelength dependence of the refractive index the Sellmeier equation can be used. It is an empirical formula that works well in describing dispersion. *Sellmeier coefficients* are often quoted instead of the refractive index in tables.

Because of dispersion, it is usually important to specify the vacuum wavelength at which a refractive index is measured. Typically, this is done at various well-defined spectralemission lines; for example, n_{D} is the refractive index at the Fraunhofer "D" line, the centre of the yellow sodium double emission at 589.29 nm wavelength.

Complex Index of Refraction and Absorption

When light passes through a medium, some part of it will always be absorbed. This can be conveniently taken into account by defining a complex index of refraction,

$$\tilde{n} = n + i\kappa.$$

Here, the real part of the refractive index n indicates the phase speed, while the imaginary part κ indicates the amount of absorption loss when the electromagnetic wave propagates through the material.

That k corresponds to absorption can be seen by inserting this refractive index into the expression for electric field of a plane electromagnetic wave travelling in the z-direction. We can do this by relating the wave number to the refractive index through $k = \frac{2\pi n}{\lambda_0}$, with λ_0 being the vacuum wavelength. With complex wave number $\tilde{k}$ and refractive index $n + i\kappa$this can be inserted into the plane wave expression as

$$\boldsymbol{E}(z,t) = \mathrm{Re}(\boldsymbol{E}_0 e^{i(\tilde{k}z-\omega t)}) = \mathrm{Re}(\boldsymbol{E}_0 e^{i(2\pi(n+i\kappa)z/\lambda_0-\omega t)}) = e^{-2\pi\kappa z/\lambda_0}\mathrm{Re}(\boldsymbol{E}_0$$

Here we see that κ gives an exponential decay, as expected from the Beer–Lambert law. k is often called the extinction coefficient in physics although this has a different definition within chemistry. Both n and k are dependent on the frequency. In most circumstancesκ > 0 (light is absorbed) or k = 0 (light travels forever without loss). In special situations, especially in the gain medium of lasers, it is also possible that k < 0, corresponding to an amplification of the light.

An alternative convention uses $\tilde{n} = n - i\kappa$ instead of $\tilde{n} = n + i\kappa$, but where k > 0 still corresponds to loss. Therefore these two conventions are inconsistent and should not be confused. The difference is related to defining sinusoidal time dependence as $\mathrm{Re}(e^{-i\omega t})$ versus $\mathrm{Re}(e^{+i\omega t})$.

Dielectric loss and non-zero DC conductivity in materials cause absorption. Good dielectric materials such as glass have extremely low DC conductivity, and at low frequencies the dielectric loss is also negligible, resulting in almost no absorption (k ≈ 0). However, at higher frequencies (such as visible light), dielectric loss may increase absorption significantly, reducing the material's transparency to these frequencies.

The real and imaginary parts of the complex refractive index are related through the Kramers–Kronig relations. For example, one can determine a material's full complex refractive index as a function of wavelength from an absorption spectrum of the material.

For X-ray and extreme ultraviolet radiation the complex refractive index deviates only slightly from unity and usually has a real part smaller than 1. It is therefore normally written as $\tilde{n} = 1 - \delta + i\beta$ (or $\tilde{n} = 1 - \delta - i\beta$).

Relations to Other Quantities

Refraction

When light moves from one medium to another as in the figure to the right, it changes direction, i.e. it is refracted. If it goes from a medium with refractive index n_1 to one with refractive index n_2, with an incidence angle to the surface normal of θ_1, the transmission angle θ_2 can be calculated from Snell's law:

$$n_1 \sin \theta_1 = n_2 \sin \theta_2.$$

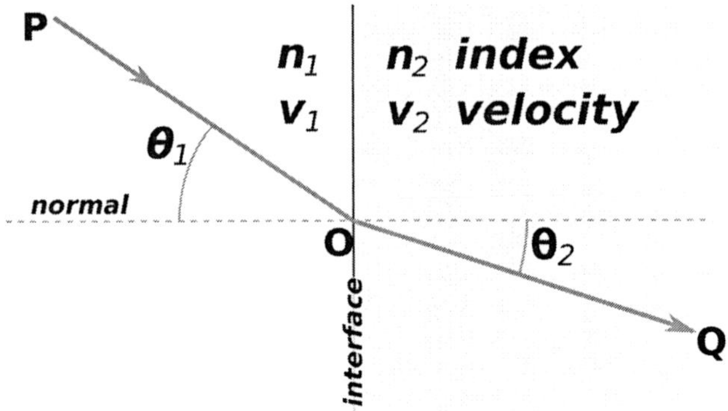

Figure 7: Refraction of light at the interface between two media of different refractive indices, with $n_2 > n_1$. Since the phase velocity is lower in the second medium ($v_2 < v_1$), the angle of refraction θ_2 is less than the angle of incidence θ_1; that is, the ray in the higher-index medium is closer to the normal.

If there is no angle θ_2 fulfilling Snell's law, i.e.

$$\frac{n_1}{n_2} \sin \theta_1 > 1,$$

the light cannot be transmitted and will instead undergo total internal reflection.

Reflectivity

Apart from the transmitted light there is also a reflected part. The reflection angle is equal to the incidence angle, and the amount of light that is reflected is determined by the reflectivity of the surface. The reflectivity can be calculated from the refractive index and the incidence angle with the Fresnel equations, which for normal incidence reduces to

$$R = \left(\frac{n_1 - n_2}{n_1 + n_2} \right)^2 .$$

For common glass in air, $n_1 = 1$ and $n_2 = 1.5$, and thus about 4% of the incident power is reflected. At other incidence angles the reflectivity will also depend on the polarisation of the incoming light. At a certain angle called Brewster's angle, p-polarised light (light with the electric field in the plane of incidence) will be totally transmitted. Brewster's angle can be calculated from the two refractive indices of the interface as

$$\theta_B = \arctan\left(\frac{n_2}{n_1}\right).$$

Lenses

The focal length of a lens is determined by its refractive index n and the radii of curvature R_1 and R_2 of its surfaces. The power of a thin lens in air is given by the Lensmaker's formula:

$$\frac{1}{f} = (n-1)\left(\frac{1}{R_1} - \frac{1}{R_2}\right).$$

Dielectric Constant

The frequency dependent dielectric constant is simply the square of the (complex) refractive index in a non-magnetic medium (one with a relative permeability of unity). The refractive index is used for optics in Fresnel equations and Snell's law; while the dielectric constant is used in Maxwell's equations and electronics.

Where $\tilde{o}$ is the complex dielectric constant with real and imaginary parts $\grave{o}_1$ and $\grave{o}_2$, and n and k are the real and imaginary parts of the refractive index, all functions of frequency:

$$\tilde{o} = \grave{o}_1 + i\grave{o}_2 = (n + i\kappa)^2.$$

Conversion between refractive index and dielectric constant is done by:

$$\grave{o}_1 = n^2 - \kappa^2$$

$$\grave{o}_2 = 2n\kappa$$

$$n = \sqrt{\frac{\sqrt{\grave{o}_1^2 + \grave{o}_2^2} + \grave{o}_1}{2}}$$

$$\kappa = \sqrt{\frac{\sqrt{\grave{o}_1^2 + \grave{o}_2^2} - \grave{o}_1}{2}}.$$

Density

In general, the refractive index of a glass increases with its density. However, there does not exist an overall linear relation between the refractive index and the density for all silicate and borosilicate glasses. A relatively high refractive index and low density can be obtained with glasses containing light metal oxides such as Li_2O and MgO, while the opposite trend is observed with glasses containing PbO and BaO as seen in the diagram at the right.

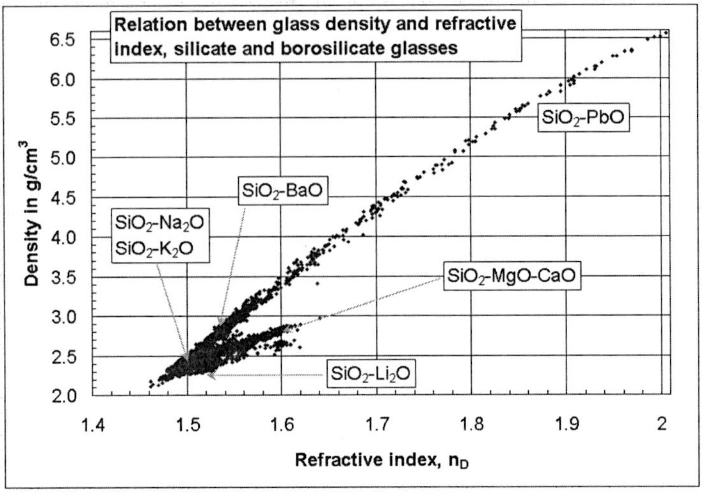

Figure 8: Relation between the refractive index and the density of silicate and borosilicate glasses.

Group Index

Sometimes, a "group speed refractive index", usually called the *group index* is defined:

$$n_g = \frac{c}{v_g}$$

where v_g is the group velocity. This value should not be confused with n, which is always defined with respect to the phase velocity. When the dispersion is small, the group velocity can be linked to the phase velocity by the relation

$$v_g = v - \lambda \frac{dv}{d\lambda}.$$

In this case the group index can thus be written in terms of the wavelength dependence of the refractive index as

$$n_g = \frac{n}{1 + \frac{\lambda}{n}\frac{dn}{d\lambda}},$$

where λ is the wavelength in the medium.

When the refractive index of a medium is known as a function of the vacuum wavelength (instead of the wavelength in the medium), the corresponding expressions for the group velocity and index are (for all values of dispersion)

$$v_g = c\left(n - \lambda_0 \frac{dn}{d\lambda_0}\right)^{-1}.$$

$$n_g = n - \lambda_0 \frac{dn}{d\lambda_0},$$

where λ_o is the wavelength in vacuum.

Other Relations

As shown in the Fizeau experiment, when light is transmitted through a moving medium, its speed relative to a stationary observer is:

$$V = \frac{c}{n} + v\left(1 - \frac{1}{n^2}\right)$$

The refractive index of a substance can be related to its polarisability with the Lorentz–Lorenz equation or to the molar refractivities of its constituents by the Gladstone–Dale relation.

Birefringence

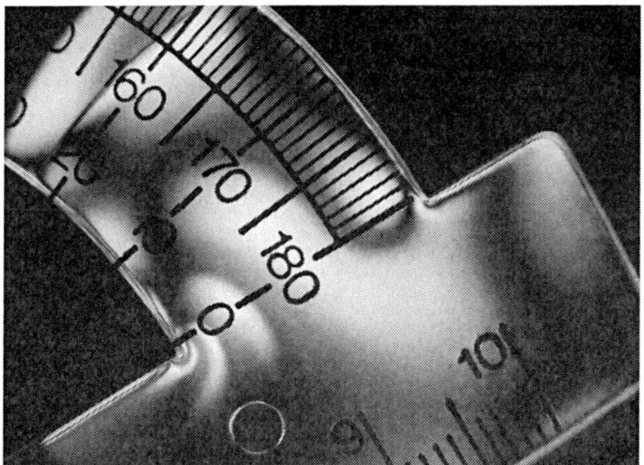

Figure 9: Birefringent materials can gives rise to colours when placed between crossed polarisers. This is the basis forphotoelasticity.

In some materials the refractive index depends on the polarisation and propagation direction of the light. This is called birefringence or optical anisotropy.

In the simplest form, uniaxial birefringence, there is only one special direction in the material. This axis is known as the optical axis of the material. Light with linear polarisation perpendicular to this axis will experience an *ordinary* refractive index n_o while light polarised in parallel will experience an *extraordinary* refractive index n_e. The birefringence of the material is the difference between these indices of refraction, $\Delta n = n_e - n_o$. Light propagating in the direction of the optical axis will not be affected by the birefringence since the refractive index will be n_o independent of polarisation.

For other propagation directions the light will split into two linearly polarised beams. For light travelling perpendicularly to the optical axis the beams will have the same direction. This can be used to change the polarisation direction of linearly polarised light or to convert between linear, circular and elliptical polarisations with waveplates.

Many crystals are naturally birefringent, but isotropic materials such as plastics and glass can also often be made birefringent by introducing a prefered direction through e.g. an external force or electric field.

This can be utilised in the determination of stresses in structures using photoelasticity. The birefringent material is then placed between crossed polarisers. A change in birefringence will alter the polarisation and thereby the fraction of light that is transmitted through the second polariser.

In the more general case of trirefringent materials described by the field of crystal optics, the *dielectric constant* is a rank-2 tensor (a 3 by 3 matrix). In this case the propagation of light cannot simply be described by refractive indices except for polarisations along principal axes.

Nonlinearity

The strong electric field of high intensity light (such as output of a laser) may cause a medium's refractive index to vary as the light passes through it, giving rise to nonlinear optics. If the index varies quadratically with the field (linearly with the intensity), it is called the optical Kerr effect and causes phenomena such as self-focusing and self-phase modulation.

If the index varies linearly with the field (which is only possible in materials that do not possess inversion symmetry), it is known as the Pockels effect.

Inhomogeneity

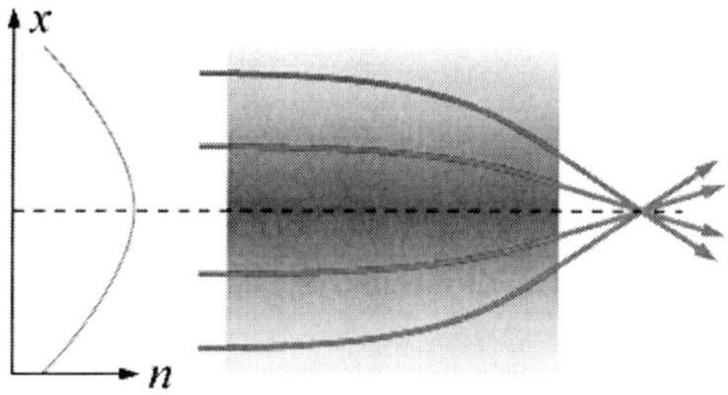

Figure 10: A gradient-index lens with a parabolic variation of refractive index (n) with radial distance (x). The lens focuses light in the same way as a conventional lens.

If the refractive index of a medium is not constant, but varies gradually with position, the material is known as a gradient-index medium and is described by gradient index optics. Light travelling through such a medium can be bent or focused, and this effect can be exploited to produce lenses, some optical fibers and other devices. Some common mirages are caused by a spatially-varying refractive index of air.

Refractive Index Measurement

Homogeneous Media

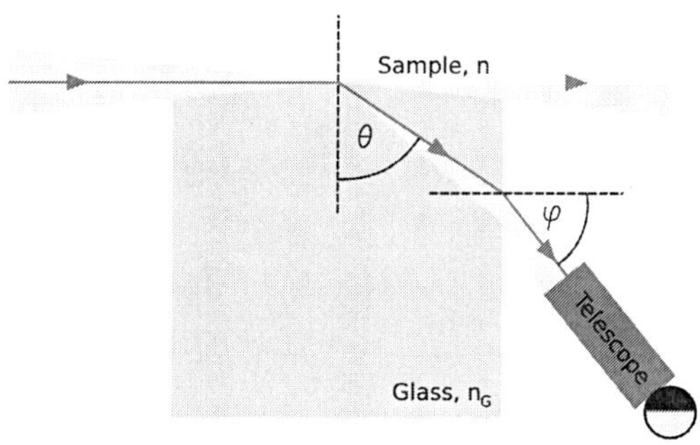

Figure 11: The principle of many refractometers.

The refractive index of liquids or solids can be measured with refractometers. They typically measure some angle of refraction or the critical angle for total internal reflection. The first laboratory refractometers sold commercially where developed by Ernst Abbe in the late 19th century. The same principles are still used today. In this instrument a thin layer of the liquid to be measured is placed between two prisms. Light is shone through the liquid at incidence angles all the way up to 90°, i.e. light rays parallel to the surface. The second prism should have an index of refraction higher than that of the liquid, so that light only enters the prism at angles smaller than the critical angle for total reflection. This angle can then be measured either by looking through a telescope, or with a digital photodetector placed in the focal plane of a lens. The refractive index n of the liquid can then be calculated from the maximum transmission angle θ as $n = n_G \sin\theta$, where n_G is the refractive index of the prism.

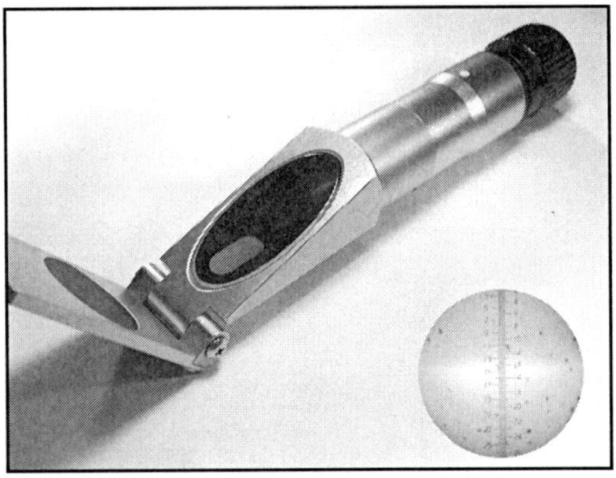

Figure 12: Photo of a handheld refractometer used to measure sugar content of fruits.

This type of devices are commonly used in chemical laboratories for identification of substances and for quality control. Handheld variants are used in agriculture by e.g. wine makers to determinesugar content in grape juice, and inline process refractometers are used in e.g. chemical and pharmaceutical industry for process control.

In gemology a different type of refractometer is used to measure index of refraction and birefringence of gemstones. The gem is placed on a high refractive index prism and illuminated from below. A high refractive index contact liquid is used to achieve optical contact between the gem and the prism. At small incidence angles most of the light will be transmitted into the gem, but at high angles total internal reflection will occur in the prism. The critical angle is normally measured by looking through a telescope.

Refractive Index Variations

To measure the spatial variation of refractive index in a sample phase-contrast imaging methods are used. These methods measure the variations in phase of the light wave exiting the sample. The phase is proportional to the optical path length the light ray has traversed, and thus gives a measure of the integral of the refractive index along the ray path. The phase cannot be measured directly at optical or higher frequencies, and therefore needs to be converted into intensity by interference with a reference beam. In the visual spectrum this is done using Zernike phase-contrast microscopy, differential interference contrast microscopy (DIC) or interferometry.

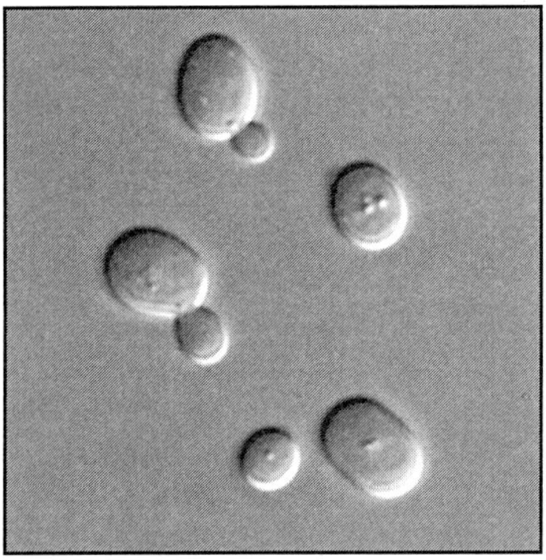

Figure 13: A differential interference contrast microscopy image of yeast cells.

Zernike phase-contrast microscopy introduces a phase shift to the low spatial frequency components of the image with a phase-shifting annulus in the Fourier plane of the sample, so that higher frequency parts of the image can interfere with the low frequency reference beam. In DIC the illumination is split up into two beams that are given different polarisations, are phase shifted differently, and are shifted transversely with slightly different amounts. After the specimen the two parts are made to interfere giving an image of the derivative of the optical path length in the direction of the difference in transverse shift. In interferometry the illumination is split up into two beams by a partially reflective mirror. One of the beams is let through the sample before they are combined to interfere and give a direct image of the phase shifts. If the optical path length variations are more than a wavelength the image will contain fringes. There exist several x-ray phase-contrast imaging techniques to determine 2D or 3D spatial distribution of refractive index of samples in the x-ray regime.

Abraham–Minkowski Controversy

In 1908, Hermann Minkowski calculated the momentum of a refracted ray, p, where E is energy of the photon, c is the speed of light in vacuum and n is the refractive index of the medium as follows:

$$p = \frac{nE}{c}.$$

In 1909, Max Abraham proposed the following formula for this calculation:

$$p = \frac{E}{nc}.$$

A 2010 study suggested that *both* equations are correct, with the Abraham version being the kinetic momentum and the Minkowski version being the canonical momentum, and claims to explain the contradicting experimental results using this interpretation.

Applications

Figure 14: Wavefronts from a point source in the context of Snell's law. The region below the gray line has a higher index of refraction, and so light travelling through it has a proportionally lower phase speed than in the region above it.

The refractive index of a material is the most important property of any optical system that uses refraction. It is used to calculate the focusing power of lenses, and the dispersive power of prisms. It can also be used as a useful tool to differentiate between different types of gemstone, due to the unique chatoyance each individual stone displays.

Since refractive index is a fundamental physical property of a substance, it is often used to identify a particular substance, confirm its purity, or measure its concentration. Refractive index is used to measure solids (glasses and gemstones), liquids, and gases. Most commonly it is used to measure the concentration of a solute in an aqueous solution. A refractometer is the instrument used to measure refractive index. For a solution of sugar, the refractive index can be used to determine the sugar content. In GPS, the index of refraction is utilised in ray-tracing to account for the radio propagation delay due to the Earth's electrically neutral atmosphere. It is also used in Satellite link design for the Computation of radiowave attenuation in the atmosphere

X-ray Fluorescence

X-ray fluorescence (XRF) is the emission of characteristic "secondary" (or fluorescent) X-rays from a material that has been excited by bombarding with high-energy X-rays orgamma rays. The phenomenon is widely used for elemental analysis and chemical analysis, particularly in the investigation of metals, glass, ceramics and building materials, and for research in geochemistry, forensic science and archaeology.

Underlying Physics

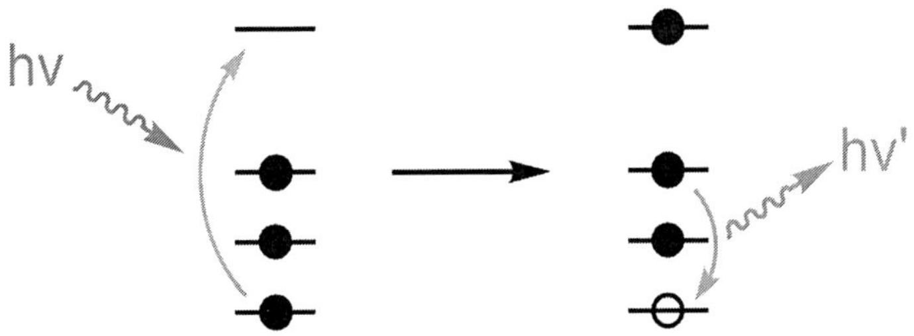

Figure 15: Physics of X-ray fluorescence, in a schematic representation.

When materials are exposed to short-wavelength X-rays or to gamma rays, ionization of their component atoms may take place. Ionization consists of the ejection of one or more electrons from the atom, and may occur if the atom is exposed to radiation with an energy greater than its ionization potential. X-rays and gamma rays can be energetic enough to expel tightly held electrons from the inner orbitals of the atom.

The removal of an electron in this way renders the electronic structure of the atom unstable, and electrons in higher orbitals "fall" into the lower orbital to fill the hole left behind. In falling, energy is released in the form of a photon, the energy of which is equal to the energy difference of the two orbitals involved. Thus, the material emits radiation, which has energy characteristic of the atoms present. The term *fluorescence* is applied to phenomena in which the absorption of radiation of a specific energy results in the re-emission of radiation of a different energy (generally lower).

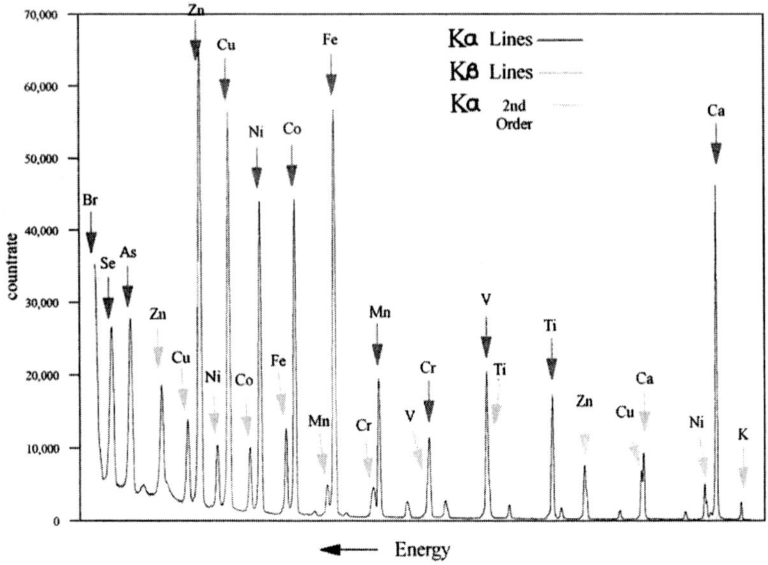

Figure 16: Typical energy dispersive XRF spectrum

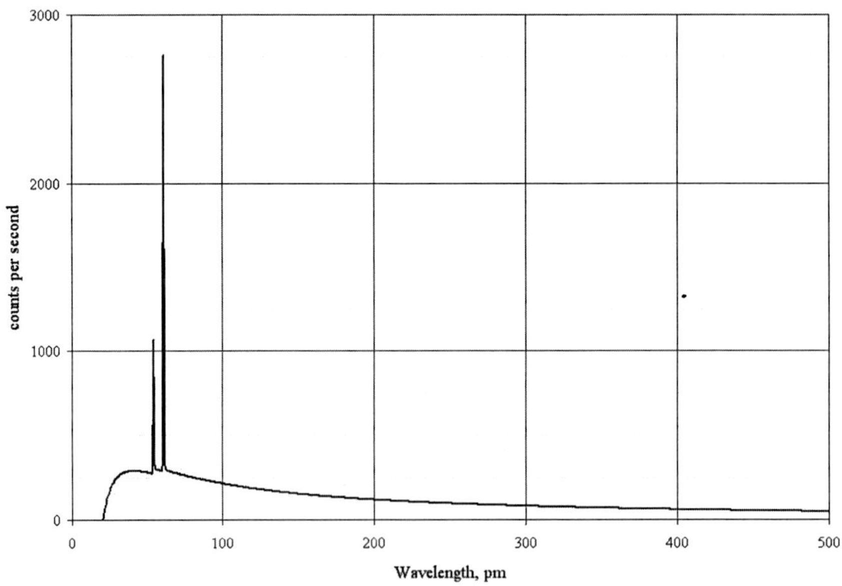

Figure 17: Spectrum of a rhodium target tube operated at 60 kV, showing continuous spectrum and K lines

Characteristic Radiation

Each element has electronic orbitals of characteristic energy. Following removal of an inner electron by an energetic photon provided by a primary radiation source, an electron from an outer shell drops into its place. There are a limited number of ways in which this can happen. The main transitions are given names: an L→K transition is traditionally called K**a**, an M→K transition is called K**b**, an M→L transition is called L**a**, and so on. Each of these transitions yields a fluorescent photon with a characteristic energy equal to the difference in energy of the initial and final orbital. The wavelength of this fluorescent radiation can be calculated from Planck's Law:

$$\lambda = h \cdot c / E$$

The fluorescent radiation can be analysed either by sorting the energies of the photons (energy-dispersive analysis) or by separating the wavelengths of the radiation (wavelength-dispersive analysis). Once sorted, the intensity of each characteristic radiation is directly related to the amount of each element in the material. This is the basis of a powerful technique in analytical chemistry.

Primary Radiation

In order to excite the atoms, a source of radiation is required, with sufficient energy to expel tightly held inner electrons. Conventional X-ray generators are most commonly used, because their output can readily be "tuned" for the application, and because higher power can be deployed relative to other techniques. However, gamma ray sources can be used without the need for an elaborate power supply, allowing an easier use in small portable instruments. When the energy source is a synchrotron or the X-rays are focused by an optic like a polycapillary, the X-ray beam can be very small and very intense. As a result, atomic information on the sub-micrometre scale can be obtained. X-ray generators in the range 20–60 kV are used, which allow excitation of a broad range of atoms. The continuous spectrum consists of "bremsstrahlung" radiation: radiation produced when high-energy electrons passing through the tube are progressively decelerated by the material of the tube anode (the "target").

Dispersion

In energy dispersive analysis, the fluorescent X-rays emitted by the material sample are directed into a solid-state detector which produces a "continuous" distribution of pulses, the voltages of which are proportional to the incoming photon energies. This signal is processed by a multichannel analyser (MCA) which produces an accumulating digital spectrum that can be processed to obtain analytical data. In wavelength dispersive analysis, the fluorescent X-rays emitted by the material sample are directed into a diffraction grating monochromator. The diffraction grating used is usually a single crystal. By varying the angle of incidence and take-off on the crystal, a single X-ray wavelength can be selected. The wavelength obtained is given by the Bragg Equation:

$$n \cdot \lambda = 2d \cdot \sin(\theta)$$

where d is the spacing of atomic layers parallel to the crystal surface.

Detection

In energy dispersive analysis, dispersion and detection are a single operation, as already mentioned above. Proportional counters or various types of solid-state detectors (PIN diode, Si(Li), Ge(Li), Silicon Drift Detector SDD) are used. They all share the same detection principle: An incoming X-ray photon ionises a large number of detector atoms with the amount of charge produced being proportional to the energy of the incoming photon. The charge is then collected and the process repeats itself for the next photon.

Detector speed is obviously critical, as all charge carriers measured have to come from the same photon to measure the photon energy correctly (peak length discrimination is used to eliminate events that seem to have been produced by two X-ray photons arriving almost simultaneously). The spectrum is then built up by dividing the energy spectrum into discrete bins and counting the number of pulses registered within each energy bin. EDXRF detector types vary in resolution, speed and the means of cooling (a low number of free charge carriers is critical in the solid state detectors): proportional counters with resolutions of several hundred eV cover the low end of the performance spectrum, followed by PIN diode detectors, while the Si(Li), Ge(Li) and Silicon Drift Detectors (SDD) occupy the high end of the performance scale.

In wavelength dispersive analysis, the single-wavelength radiation produced by the monochromator is passed into a photomultiplier, a detector similar to a Geiger counter, which counts individual photons as they pass through. The counter is a chamber containing a gas that is ionised by X-ray photons. A central electrode is charged at (typically) +1700 V with respect to the conducting chamber walls, and each photon triggers a pulse-like cascade of current across this field. The signal is amplified and transformed into an accumulating digital count. These counts are then processed to obtain analytical data.

X-ray Intensity

The fluorescence process is inefficient, and the secondary radiation is much weaker than the primary beam. Furthermore, the secondary radiation from lighter elements is of relatively low energy (long wavelength) and has low penetrating power, and is severely attenuated if the beam passes through air for any distance. Because of this, for high-performance analysis, the path from tube to sample to detector is maintained under high vacuum (around 10 Pa residual pressure). This means in practice that most of the working parts of the instrument have to be located in a large vacuum chamber. The problems of maintaining moving parts in vacuum, and of rapidly introducing and withdrawing the sample without losing vacuum, pose major challenges for the design of the instrument. For less demanding applications, or when the sample is damaged by a vacuum (e.g. a volatile sample), a helium-swept X-ray chamber can be substituted, with some loss of low-Z (Z = atomic number) intensities.

Chemical Analysis

The use of a primary X-ray beam to excite fluorescent radiation from the sample was first proposed by Glocker and Schreiber in 1928. Today, the method is used as a non-destructive analytical technique, and as a process control tool in many extractive and processing industries. In principle, the lightest element that can be analysed is beryllium (Z = 4), but due to instrumental limitations and low X-ray yields for the light elements, it is often difficult to quantify elements lighter than sodium (Z = 11), unless background corrections and very comprehensive inter-element corrections are made.

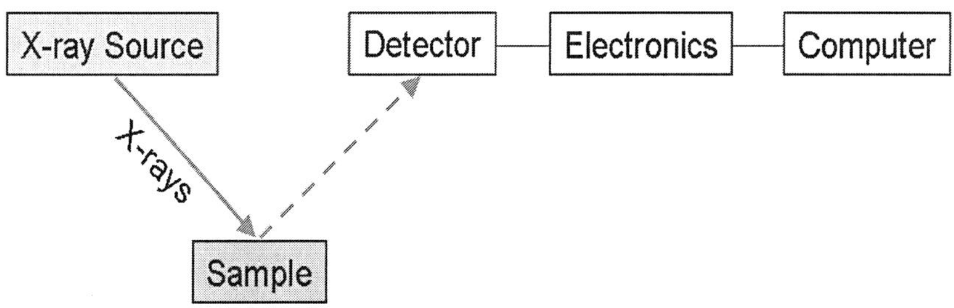

Figure 18: Schematic arrangement of EDX spectrometer

Energy Dispersive Spectrometry

In energy dispersive spectrometers (EDX or EDS), the detector allows the determination of the energy of the photon when it is detected. Detectors historically have been based on silicon semiconductors, in the form of lithium-drifted silicon crystals, or high-purity silicon wafers.

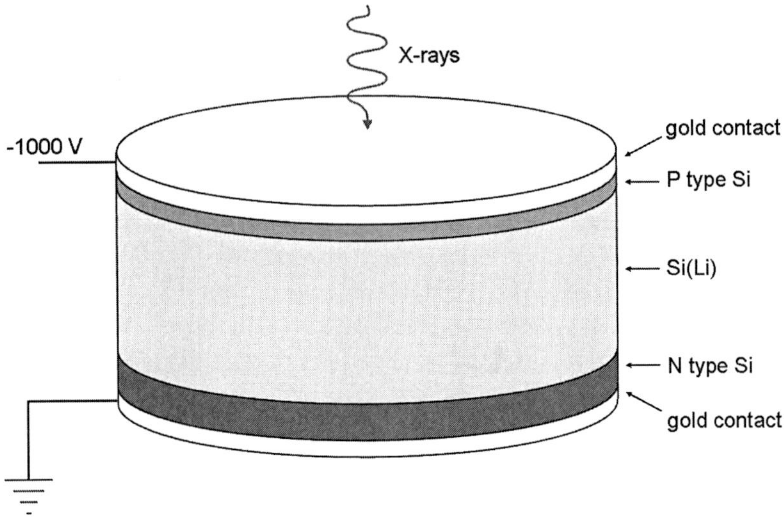

Figure 19: Schematic form of a Si(Li) detector

Si(Li) Detectors

These consist essentially of a 3–5 mm thick silicon junction type p-i-n diode (same as PIN diode) with a bias of "1000 V across it. The lithium-drifted centre part forms the non-conducting i-layer, where Li compensates the residual acceptors which would otherwise make the layer p-type. When an X-ray photon passes through, it causes a swarm of electron-hole pairs to form, and this causes a voltage pulse. To obtain sufficiently low conductivity, the detector must be maintained at low temperature, and liquid-nitrogen must be used for the best resolution. With some loss of resolution, the much more convenient Peltier cooling can be employed.

Wafer Detectors

More recently, high-purity silicon wafers with low conductivity have become routinely available. Cooled by the Peltier effect, this provides a cheap and convenient detector, although the liquid nitrogen cooled Si(Li) detector still has the best resolution (i.e. ability to distinguish different photon energies).

Amplifiers

The pulses generated by the detector are processed by pulse-shaping amplifiers. It takes time for the amplifier to shape the pulse for optimum resolution, and there is therefore a trade-off between resolution and count-rate: long processing time for good resolution results in "pulse pile-up" in which the pulses from successive photons overlap. Multi-photon events are, however, typically more drawn out in time (photons did not arrive exactly at the same time) than single photon events and pulse-length discrimination can thus be used to filter most of these out.

Even so, a small number of pile-up peaks will remain and pile-up correction should be built into the software in applications that require trace analysis. To make the most efficient use of the detector, the tube current should be reduced to keep multi-photon events (before discrimination) at a reasonable level, e.g. 5–20%.

Processing

Considerable computer power is dedicated to correcting for pulse-pile up and for extraction of data from poorly resolved spectra. These elaborate correction processes tend to be based on empirical relationships that may change with time, so that continuous vigilance is required in order to obtain chemical data of adequate precision.

Usage

EDX spectrometers are superior to WDX spectrometers in that they are smaller, simpler in design and have fewer engineered parts. They can also use miniature X-ray tubes or gamma sources. This makes them cheaper and allows miniaturisation and portability. This type of instrument is commonly used for portable quality control screening applications, such as testing toys for Lead (Pb) content, sorting scrap metals, and measuring the lead content of residential paint. On the other hand, the low resolution and problems with low count rate and long dead-time makes them inferior for high-precision analysis. They are, however, very effective for high-speed, multi-elemental analysis. Field Portable XRF analysers currently on the market weigh less than 2 kg, and have limits of detection on the order of 2 parts per million of Lead (Pb) in pure sand.

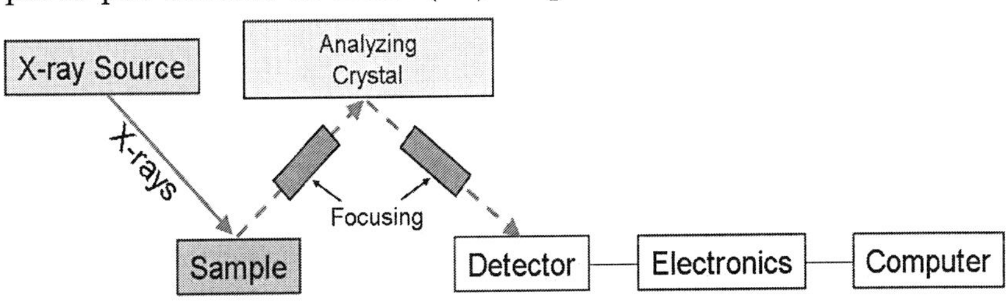

Figure 20: Schematic arrangement of wavelength dispersive spectrometer

Figure 21: Chemist operates a goniometer used for X-ray fluorescence analysis of individual grains of mineral specimens, U.S. Geological Survey, 1958.

Wavelength Dispersive Spectrometry

In wavelength dispersive spectrometers (WDX or WDS), the photons are separated by diffraction on a single crystal before being detected. Although wavelength dispersive spectrometers are occasionally used to scan a wide range of wavelengths, producing a spectrum plot as in EDS, they are usually set up to make measurements only at the wavelength of the emission lines of the elements of interest. This is achieved in two different ways:

"Simultaneous" spectrometers have a number of "channels" dedicated to analysis of a single element, each consisting of a fixed-geometry crystal monochromator, a detector, and processing electronics. This allows a number of elements to be measured simultaneously, and in the case of high-powered instruments, complete high-precision analyses can be obtained in under 30 s.

Another advantage of this arrangement is that the fixed-geometry monochromators have no continuously moving parts, and so are very reliable. Reliability is important in production environments where instruments are expected to work without interruption for months at a time. Disadvantages of simultaneous spectrometers include relatively high cost for complex analyses, since each channel used is expensive.

The number of elements that can be measured is limited to 15–20, because of space limitations on the number of monochromators that can be crowded around the fluorescing sample. The need to accommodate multiple monochromators means that a rather open arrangement around the sample is required, leading to relatively long tube-sample-crystal distances, which leads to lower detected intensities and more scattering. The instrument is inflexible, because if a new element is to be measured, a new measurement channel has to be bought and installed.

"Sequential" spectrometers have a single variable-geometry monochromator (but usually with an arrangement for selecting from a choice of crystals), a single detector assembly (but usually with more than one detector arranged in tandem), and a single electronic pack. The instrument is programmed to move through a sequence of wavelengths, in each case selecting the appropriate X-ray tube power, the appropriate crystal, and the appropriate detector arrangement. The length of the measurement program is essentially unlimited, so this arrangement is very flexible. Because there is only one monochromator, the tube-sample-crystal distances can be kept very short, resulting in minimal loss of detected intensity. The obvious disadvantage is relatively long analysis time, particularly when many elements are being analysed, not only because the elements are measured in sequence, but also because a certain amount of time is taken in readjusting the monochromator geometry between measurements. Furthermore, the frenzied activity of the monochromator during an analysis program is a challenge for mechanical reliability. However, modern sequential instruments can achieve reliability almost as good as that of simultaneous instruments, even in continuous-usage applications.

Sample Presentation

In order to keep the geometry of the tube-sample-detector assembly constant, the sample is normally prepared as a flat disc, typically of diameter 20–50 mm. This is located at a standardised, small distance from the tube window. Because the X-ray intensity follows an inverse-square law, the tolerances for this placement and for the flatness of the surface must be very tight in order to maintain a repeatable X-ray flux. Ways of obtaining sample discs vary: metals may be machined to shape, minerals may be finely ground and pressed into a tablet, and glasses may be cast to the required shape. A further reason for obtaining a flat and representative sample surface is that the secondary X-rays from lighter elements often only emit from the top few micrometres of the sample. In order to further reduce the effect of surface irregularities, the sample is usually spun at 5–20 rpm. It is necessary to ensure that the sample is sufficiently thick to absorb the entire primary beam. For higher-Z materials, a few millimetres thickness is adequate, but for a light-element matrix such as coal, a thickness of 30–40 mm is needed.

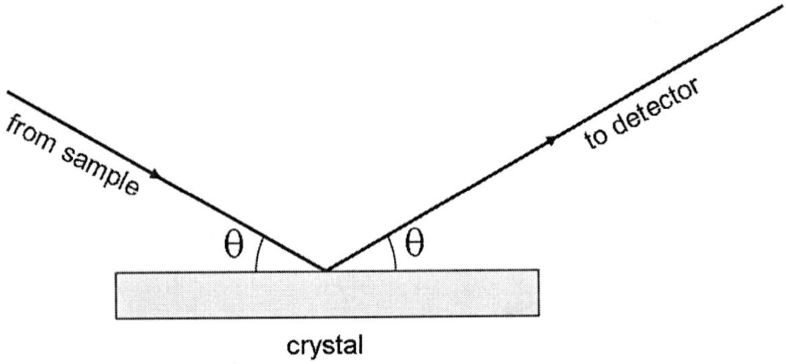

Figure 22: Bragg diffraction condition

Monochromators

The common feature of monochromators is the maintenance of a symmetrical geometry between the sample, the crystal and the detector. In this geometry the Bragg diffraction condition is obtained.

The X-ray emission lines are very narrow, so the angles must be defined with considerable precision. This is achieved in two ways:

Flat crystal with Soller collimators: The Soller collimator is a stack of parallel metal plates, spaced a few tenths of a millimetre apart. To improve angle resolution, one must lengthen the collimator, and/or reduce the plate spacing. This arrangement has the advantage of simplicity and relatively low cost, but the collimators reduce intensity and increase scattering, and reduce the area of sample and crystal that can be "seen". The simplicity of the geometry is especially useful for variable-geometry monochromators.

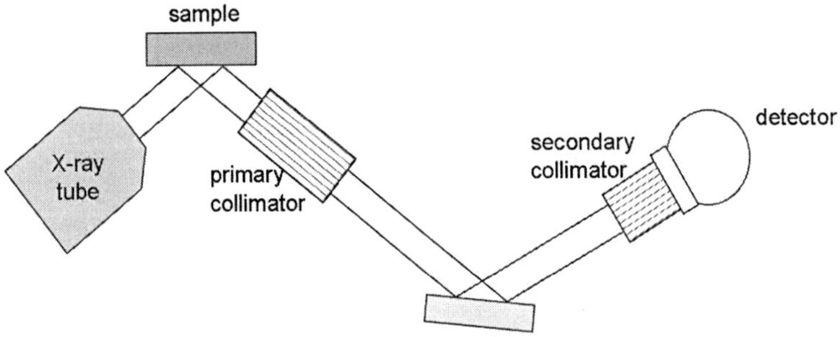

Figure 23: Flat crystal with Soller collimators

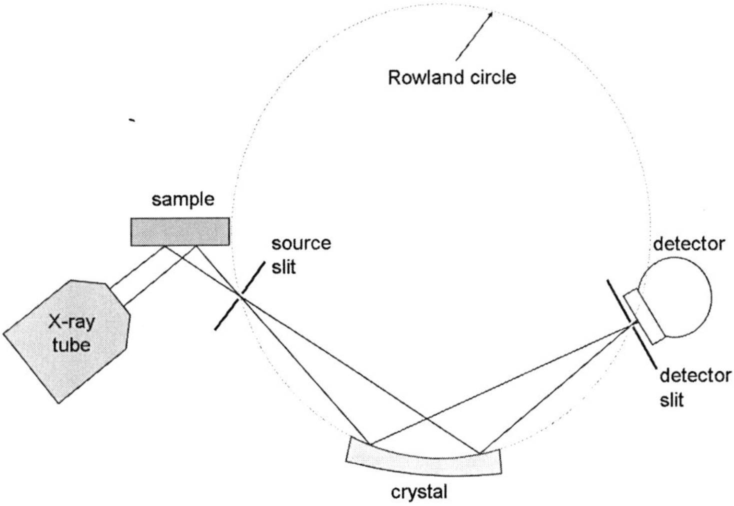

Figure 24: Curved crystal with slits

Curved crystal with slits: The Rowland circle geometry ensures that the slits are both in focus, but in order for the Bragg condition to be met at all points, the crystal must first be bent to a radius of 2R (where R is the radius of the Rowland circle), then ground to a radius of R. This arrangement allows higher intensities (typically 8-fold) with higher resolution (typically 4-fold) and lower background. However, the mechanics of keeping Rowland circle geometry in a variable-angle monochromator is extremely difficult. In the case of fixed-angle monochromators (for use in simultaneous spectrometers), crystals bent to a logarithmic spiral shape give the best focusing performance. The manufacture of curved crystals to acceptable tolerances increases their price considerably.

Crystals

The desirable characteristics of a diffraction crystal are:

- High diffraction intensity
- High dispersion
- Narrow diffracted peak width
- High peak-to-background
- Absence of interfering elements
- Low thermal coefficient of expansion
- Stability in air and on exposure to X-rays
- Ready availability
- Low cost.

Crystals with simple structure tend to give the best diffraction performance. Crystals containing heavy atoms can diffract well, but also fluoresce themselves, causing interference. Crystals that are water-soluble, volatile or organic tend to give poor stability.

Commonly used crystal materials include LiF (lithium fluoride), ADP (ammonium dihydrogen phosphate), Ge (germanium), graphite, InSb (indium antimonide), PE (*tetrakis*-(hydroxymethyl)-methane: penta-erythritol), KAP (potassium hydrogen phthalate), RbAP (rubidium hydrogen phthalate) and TlAP (thallium(I) hydrogen phthalate). In addition, there is an increasing use of "layered synthetic microstructures", which are "sandwich" structured materials comprising successive thick layers of low atomic number matrix, and monatomic layers of a heavy element. These can in principle be custom-manufactured to diffract any desired long wavelength, and are used extensively for elements in the range Li to Mg.

Table 1: Properties of commonly used crystals

Material	*plane*	*d (nm)*	*min λ (nm)*	*max λ (nm)*	*intensity*	*thermal expansion*	*durability*
LiF	200	0.2014	0.053	0.379	+++++	+++	+++
LiF	220	0.1424	0.037	0.268	+++	++	+++
LiF	420	0.0901	0.024	0.169	++	++	+++
ADP	101	0.5320	0.139	1.000	+	++	++
Ge	111	0.3266	0.085	0.614	+++	+	+++
graphite	001	0.3354	0.088	0.630	++++	+	+++
InSb	111	0.3740	0.098	0.703	++++	+	+++
PE	002	0.4371	0.114	0.821	+++	+++++	+
KAP	1010	1.325	0.346	2.490	++	++	++
RbAP	1010	1.305	0.341	2.453	++	++	++
Si	111	0.3135	0.082	0.589	++	+	+++
TlAP	1010	1.295	0.338	2.434	+++	++	++
YB_{66}	400	0.586					
6 nm LSM	-	6.00	1.566	11.276	+++	+	++

Detectors

Detectors used for wavelength dispersive spectrometry need to have high pulse processing speeds in order to cope with the very high photon count rates that can be obtained. In addition, they need sufficient energy resolution to allow filtering-out of background noise and spurious photons from the primary beam or from crystal fluorescence. There are four common types of detector:

- gas flow proportional counters
- sealed gas detectors
- scintillation counters
- semiconductor detectors.

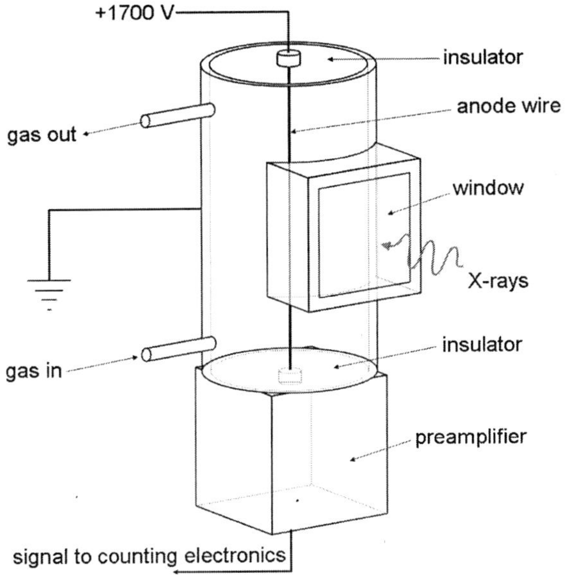

Figure 25: Arrangement of gas flow proportional counter

Gas flow proportional counters are used mainly for detection of longer wavelengths. Gas flows through it continuously. Where there are multiple detectors, the gas is passed through them in series, then led to waste. The gas is usually 90% argon, 10% methane ("P10"), although the argon may be replaced with neon or helium where very long wavelengths (over 5 nm) are to be detected. The argon is ionised by incoming X-ray photons, and the electric field multiplies this charge into a measurable pulse.

The methane suppresses the formation of fluorescent photons caused by recombination of the argon ions with stray electrons. The anode wire is typically tungsten or nichrome of 20–60 ìm diameter. Since the pulse strength obtained is essentially proportional to the ratio of the detector chamber diameter to the wire diameter, a fine wire is needed, but it must also be strong enough to be maintained under tension so that it remains precisely straight and concentric with the detector. The window needs to be conductive, thin enough to transmit the X-rays effectively, but thick and strong enough to minimise diffusion of the detector gas into the high vacuum of the monochromator chamber. Materials often used are beryllium metal, aluminised PET film and aluminised polypropylene.

Ultra-thin windows (down to 1 ìm) for use with low-penetration long wavelengths are very expensive. The pulses are sorted electronically by "pulse height selection" in order to isolate those pulses deriving from the secondary X-ray photons being counted.

Sealed gas detectors are similar to the gas flow proportional counter, except that the gas does not flow through it. The gas is usually krypton or xenon at a few atmospheres pressure. They are applied usually to wavelengths in the 0.15–0.6 nm range. They are applicable in principle to longer wavelengths, but are limited by the problem of manufacturing a thin window capable of withstanding the high pressure difference.

Scintillation counters consist of a scintillating crystal (typically of sodium iodide doped with thallium) attached to a photomultiplier. The crystal produces a group of scintillations for each photon absorbed, the number being proportional to the photon energy. This translates into a pulse from the photomultiplier of voltage proportional to the photon energy.

The crystal must be protected with a relatively thick aluminium/ beryllium foil window, which limits the use of the detector to wavelengths below 0.25 nm. Scintillation counters are often connected in series with a gas flow proportional counter: the latter is provided with an outlet window opposite the inlet, to which the scintillation counter is attached. This arrangement is particularly used in sequential spectrometers.

Semiconductor detectors can be used in theory, and their applications are increasing as their technology improves, but historically their use for WDX has been restricted by their slow response.

Extracting Analytical Results

At first sight, the translation of X-ray photon count-rates into elemental concentrations would appear to be straightforward: WDX separates the X-ray lines efficiently, and the rate of *generation* of secondary photons is proportional to the element concentration. However, the number of photons *leaving the sample* is also affected by the physical properties of the sample: so-called "matrix effects". These fall broadly into three categories:

- X-ray absorption
- X-ray enhancement
- sample macroscopic effects

All elements *absorb* X-rays to some extent. Each element has a characteristic absorption spectrum which consists of a "saw-tooth" succession of fringes, each step-change of which has wavelength close to an emission line of the element. Absorption attenuates the secondary X-rays leaving the sample. For example, the mass absorption coefficient of silicon at the wavelength of the aluminium K**a** line is 50 m^2/kg, whereas that of iron is 377 m^2/kg.

This means that a given concentration of aluminium in a matrix of iron gives only one seventh of the count rate compared with the same concentration of aluminium in a silicon matrix. Fortunately, mass absorption coefficients are well known and can be calculated. However, to calculate the absorption for a multi-element sample, the composition must be known. For analysis of an unknown sample, an iterative procedure is therefore used. It will be noted that, to derive the mass absorption accurately, data for the concentration of elements not measured by XRF may be needed, and various strategies are employed to estimate these. As an example, in cement analysis, the concentration of oxygen (which is not measured) is calculated by assuming that all other elements are present as standard oxides.

Enhancement occurs where the secondary X-rays emitted by a heavier element are sufficiently energetic to stimulate additional secondary emission from a lighter element. This phenomenon can also be modelled, and corrections can be made provided that the full matrix composition can be deduced.

Sample macroscopic effects consist of effects of inhomogeneities of the sample, and unrepresentative conditions at its surface. Samples are ideally homogeneous and isotropic, but they often deviate from this ideal. Mixtures of multiple crystalline components in mineral powders can result in absorption effects that deviate from those calculable from theory.

When a powder is pressed into a tablet, the finer minerals concentrate at the surface. Spherical grains tend to migrate to the surface more than do angular grains. In machined metals, the softer components of an alloy tend to smear across the surface. Considerable care and ingenuity are required to minimise these effects. Because they are artifacts of the method of sample preparation, these effects can not be compensated by theoretical corrections, and must be "calibrated in". This means that the calibration materials and the unknowns must be compositionally and mechanically similar, and a given calibration is applicable only to a limited range of materials. Glasses most closely approach the ideal of homogeneity and isotropy, and for accurate work, minerals are usually prepared by dissolving them in a borate glass, and casting them into a flat disc or "bead". Prepared in this form, a virtually universal calibration is applicable.

Further corrections that are often employed include background correction and line overlap correction. The background signal in an XRF spectrum derives primarily from scattering of primary beam photons by the sample surface. Scattering varies with the sample mass absorption, being greatest when mean atomic number is low. When measuring trace amounts of an element, or when measuring on a variable light matrix, background correction becomes necessary. This is really only feasible on a sequential spectrometer. Line overlap is a common problem, bearing in mind that the spectrum of a complex mineral can contain several hundred measurable lines. Sometimes it can be overcome by measuring a less-intense, but overlap-free line, but in certain instances a correction is inevitable. For instance, the K**a** is the only usable line for measuring sodium, and it overlaps the zinc L**b** (L_2-M_4) line. Thus zinc, if present, must be analysed in order to properly correct the sodium value.

Other Spectroscopic Methods Using the Same Principle

It is also possible to create a characteristic secondary X-ray emission using other incident radiation to excite the sample:

- electron beam: electron microprobe (or Castaing microprobe);
- ion beam: particle induced X-ray emission (PIXE).

When radiated by an X-ray beam, the sample also emits other radiations that can be used for analysis:

- electrons ejected by the photoelectric effect: X-ray photoelectron spectroscopy (XPS), also called electron spectroscopy for chemical analysis (ESCA).

The de-excitation also ejects Auger electrons, but Auger electron spectroscopy (AES) normally uses an electron beam as the probe.

Confocal microscopy X-ray fluorescence imaging is a newer technique that allow control over depth, in addition to horizontal and vertical aiming, for example, when analysing buried layers in a painting.

Instrument Qualification

A 2001 review, addresses the application of portable instrumentation from QA/QC perspectives. It provides a guide to the development of a set of SOPs if regulatory compliance guidelines are not available.

Emission Spectrum

The emission spectrum of a chemical element or chemical compound is the spectrum of frequencies of electromagnetic radiation emitted by the element's atoms or the compound's molecules when they are returned to a lower energy state.

Each element's emission spectrum is unique. Therefore, spectroscopy can be used to identify the elements in matter of unknown composition. Similarly, the emission spectra of molecules can be used in chemical analysis of substances.

Emission (Light)

In physics, emission is the process by which a higher energy quantum mechanical state of a particle becomes converted to a lower one through the emission of a photon, resulting in the production of light. The frequency of light emitted is a function of the energy of the transition. Since energy must be conserved, the energy difference between the two states equals the energy carried off by the photon. The energy states of the transitions can lead to emissions over a very large range of frequencies. For example, visible light is emitted by the coupling of electronic states in atoms and molecules (then the phenomenon is called fluorescence or phosphorescence). On the other hand, nuclear shell transitions can emit high energy gamma rays, while nuclear spin transitions emit low energy radiowaves.

The emittance of an object quantifies how much light is emitted by it. This may be related to other properties of the object through the Stefan–Boltzmann law. For most substances, the amount of emission varies with the temperature and the spectroscopic composition of the object, leading to the appearance of color temperature and emission lines. Precise measurements at many wavelengths allow the identification of a substance via emission spectroscopy.

Emission of radiation is typically described using semi-classical quantum mechanics: the particle's energy levels and spacings are determined from quantum mechanics, and light is treated as an oscillating electric field that can drive a transition if it is in resonance with the system's natural frequency. The quantum mechanics problem is treated using time-dependent perturbation theory and leads to the general result known as Fermi's golden rule. The description has been superseded by quantum electrodynamics, although the semi-classical version continues to be more useful in most cases.

Origins

When the electrons in the atom are excited, for example by being heated, the additional energy pushes the electrons to higher energy orbitals. When the electrons fall back down and leave the excited state, energy is re-emitted in the form of a photon. The wavelength (or equivalently, frequency) of the photon is determined by the difference in energy between the two states. These emitted photons form the element's emission spectrum.

The fact that only certain colours appear in an element's atomic emission spectrum means that only certain frequencies of light are emitted. Each of these frequencies are related to energy by the formula:

$$E_{\text{photon}} = h\,\nu,$$

where E is the energy of the photon, ν is its frequency, and h is Planck's constant. This concludes that only photons having certain energies are emitted by the atom. The principle of the atomic emission spectrum explains the varied colours inneon signs, as well as chemical flame test results mentioned above.

The frequencies of light that an atom can emit are dependent on states the electrons can be in. When excited, an electron moves to a higher energy level/orbital. When the electron falls back to its ground level the light is emitted.

Radiation From Molecules

As well as the electronic transitions discussed above, the energy of a molecule can also change via rotational, vibrational, and vibronic (combined vibrational and electronic) transitions. These energy transitions often lead to closely spaced groups of many different spectral lines, known as spectral bands. Unresolved band spectra may appear as a spectral continuum.

Molecular emission is the mechanism behind the sulfur lamp and the deuterium arc lamp.

Emission Spectroscopy

Light consists of electromagnetic radiation of different wavelengths. Therefore, when the elements or their compounds are heated either on a flame or by an electric arc they emit energy in form of light. Analysis of this light, with the help of aspectroscope gives us a discontinuous spectrum. A spectroscope or a spectrometer is an instrument which is used for separating the components of light, which have different wavelengths. The spectrum appears in a series of lines called the line spectrum. This line spectrum is also called the Atomic Spectrum because it originates in the element. Each element has a different atomic spectrum. The production of line spectra by the atoms of an element indicate that an atom can radiate only a certain amount of energy. This leads to the conclusion that electrons cannot have just any amount of energy but only a certain amount of energy.

The emission spectrum can be used to determine the composition of a material, since it is different for each element of the periodic table. One example is astronomical spectroscopy: identifying the composition of stars by analysing the received light. The emission spectrum characteristics of some elements are plainly visible to the naked eye when these elements are heated. For example, when platinum wire is dipped into a strontium nitrate solution and then inserted into a flame, the strontium atoms emit a red colour. Similarly, when copper is inserted into a flame, the flame becomes green. These definite characteristics allow elements to be identified by their atomic emission spectrum. Not all lights emitted by the spectrum are viewable to the naked eye, it also includes ultra violet rays and infra red lighting, an emission is formed when an excited gas is viewed directly through a spectroscope.

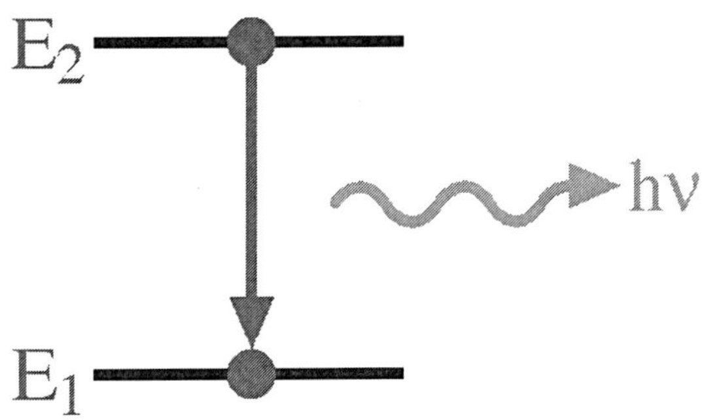

Figure 26: Schematic diagram of spontaneous emission

Emission spectroscopy is a spectroscopic technique which examines the wavelengths of photons emitted by atoms or molecules during their transition from an excited state to a lower energy state. Each element emits a characteristic set of discrete wavelengths according to its electronic structure, by observing these wavelengths the elemental composition of the sample can be determined. Emission spectroscopy developed in the late 19th century and efforts in theoretical explanation of atomic emission spectra eventually led to quantum mechanics.

There are many ways in which atoms can be brought to an excited state. Interaction with electromagnetic radiation is used in fluorescence spectroscopy, protons or other heavier particles in Particle-Induced X-ray Emission and electrons or X-ray photons in Energy-dispersive X-ray spectroscopy or X-ray fluorescence.

The simplest method is to heat the sample to a high temperature, after which the excitations are produced by collisions between the sample atoms. This method is used in flame emission spectroscopy, and it was also the method used by Anders Jonas Engström when he discovered the phenomenon of discrete emission lines in 1850s.

Although the emission lines are caused by a transition between quantised energy states and may at first look very sharp, they do have a finite width, i.e. they are composed of more than one wavelength of light. This spectral line broadening has many different causes.

Emission spectroscopy is often referred to as optical emission spectroscopy, due to the light nature of what is being emitted.

History

Emission lines from hot gases were first discovered by Engström, and the technique was further developed by David Alter, Gustav Kirchhoff and Robert Bunsen.

Experimental Technique in Flame Emission Spectroscopy

The solution containing the relevant substance to be analysed is drawn into the burner and dispersed into the flame as a fine spray. The solvent evaporates first, leaving finely divided solid particles which move to the hottest region of the flame where gaseous atoms and ions are produced. Here electrons are excited as described above. It is common for a monochromator to be used to allow for easy detection.

On a simple level, flame emission spectroscopy can be observed using just a Flame and samples of metal salts. This method of Qualitative analysis is called a Flame test.

For example, Sodium salts placed in the flame will glow yellow from Sodium ions, while Strontium (used in road flares) ions colour it red. Copper wire will create a blue coloured flame, however in the presence of Chloride gives green (molecular contribution by CuCl).

Emission Coefficient

Emission coefficient is a coefficient in the power output per unit time of an electromagnetic source, a calculated value in physics. It is also used as a measure of environmental emissions (by mass) per MWh of electricity generated, see: Emission factor.

Scattering of Light

In Thomson scattering a charged particle emits radiation under incident light. The particle may be an ordinary atomic electron, so emission coefficients have practical applications.

The values of X in Thomson scattering can be predicted from incident flux, the density of the charged particles and their Thomson differential cross section (area/solid angle).

Spontaneous Emission

A warm body emitting photons has a monochromatic emission coefficient relating to its temperature and total power radiation. This is sometimes called the second "Einstein coefficient", and can be deduced from quantum mechanical theory.

Energy Spectrum

An energy spectrum is a distribution energy among a large assemblage of particles. It is a statistical representation of the wave energy as a function of the wave frequency, and an empirical estimator of the spectral function. For any given value of energy, it determines how many of the particles have that much energy.

The particles may be atoms, photons or a flux of elementary particles.

The Schrödinger equation and a set of boundary conditions form an eigenvalue problem. A possible value (E) is called an eigenenergy. A non-zero solution of the wave function is called an eigenenergy state, or simply an eigenstate. The set of eigenvalues $\{E_j\}$ is called the energy spectrum of the particle.

The electromagnetic spectrum can also be represented as the distribution of electromagnetic radiation according to energy. The relationship among the wavelength, the frequency (usually denoted by Greek "ν"), and the energy E are:

$$E = h\nu = \frac{hc}{\lambda}$$

where c is the speed of light and h is Planck's Constant.

An example of an energy spectrum in the physical domain is ocean waves breaking on the shore. For any given interval of time it can be observed that some of the waves are larger than others. Plotting the number of waves against the amplitude (height) for the interval will yield the energy spectrum of the set.

Optical Spectroscopy and Astrophysics Application

The quantity plotted, energy units, is the wavelength times the energy per unit wavelength and thus accurately represents the amount of energy at any wavelength. The energy per unit wavelength and the energy per unit frequency peak at significantly different wavelengths due the reciprocal relation between frequency and wavelength. Using energy units avoids this problem, since (wavelength * flux per unit wavelength) = (frequency * flux per unit frequency).

Some modern spectrophotometers, such as the Perkin Elmer 950, include an energy scan option. This is additionally useful in cases where a reference cell is not practical or when absorbance / transmittance is off-scale

Atomic Spectral Line

In physics, atomic spectral lines are of two types:

- An emission line is formed when an electron makes a transition from a particular discrete energy level of an atom, to a lower energy state, emitting a photon of a particular energy and wavelength. A spectrum of many such photons will show an emission spike at the wavelength associated with these photons.
- An absorption line is formed when an electron makes a transition from a lower to a higher discrete energy state, with a photon being absorbed in the process. These absorbed photons generally come from background continuum radiation and a spectrum will show a drop in the continuum radiation at the wavelength associated with the absorbed photons.

The two states must be bound states in which the electron is bound to the atom, so the transition is sometimes referred to as a "bound–bound" transition, as opposed to a transition in which the electron is ejected out of the atom completely ("bound–free" transition) into a continuum state, leaving an ionized atom, and generating continuum radiation.

A photon with an energy equal to the energy difference between the levels is released or absorbed in the process. The frequency ν at which the spectral line occurs is related to the photon energy *E*by Planck's law $E = h\nu$ where h is Planck's constant.

Emission and Absorption Coefficients

An atomic spectral line refers to emission and absorption events in a gas in which n_2 is the density of atoms in the upper energy state for the line, and n_1 is the density of atoms in the lower energy state for the line.

The emission of atomic line radiation may be described by an emission coefficient ε with units of energy/time/volume/solid angle. ε dt dV dΩ is then the energy emitted by a volume element dV in time dt into solid angle dΩ. For atomic line radiation:

$$\grave{o} = \frac{h\nu}{4\pi} n_2 A_{21}$$

where A_{21} is the Einstein coefficient for spontaneous emission, which is fixed by the intrinsic properties of the relevant atom for the two relevant energy levels.

The absorption of atomic line radiation may be described by an absorption coefficient κ with units of 1/length. The expression κ' dx gives the fraction of intensity absorbed for a light beam at frequency ν while travelling distance dx. The absorption coefficient is given by:

$$\kappa' = \frac{h\nu}{4\pi}\,(n_1 B_{12} - n_2 B_{21})$$

where B_{12} and B_{21} are the Einstein coefficients for photo absorption and induced emission respectively. Like the coefficient A_{21}, these are also fixed by the intrinsic properties of the relevant atom for the two relevant energy levels. For thermodynamics and for the application of Kirchhoff's law, it is necessary that the total absorption be expressed as the algebraic sum of two components, described respectively by B_{12} and B_{21}, which may be regarded as positive and negative absorption, respectively the photo absorption and what is commonly called stimulated or induced emission.

The above equations have ignored the influence of the spectral line shape. To be accurate, the above equations need to be multiplied by the (normalised) spectral line shape, in which case the units will change to include a 1/Hz term.

The number densities n_2 and n_1 and the Einstein coefficients provide sufficient information to determine the absorption and emission rates, without further need for information about the physical state of the gas or about the local radiation field. Thus the above absorption and emission coefficients work for any state of the gas, so for equilibrium states and even for laser conditions.

In contrast to the Einstein coefficients, the number densities, however, are not set simply by the intrinsic properties of the relevant atom.

Equilibrium Conditions

The number densities n_2 and n_1 are set by the physical state of the gas in which the spectral line occurs, including the local specific radiative intensity. When that state is either one of strict thermodynamic equilibrium, or one of so-called 'local thermodynamic equilibrium', then the distribution of atomic states of excitation (which includes n_2 and n_1) determines the rates of atomic emissions and absorptions to be such that Kirchhoff's law of equality of radiative absorptivity and emissivity shall hold. In strict thermodynamic equilibrium, the radiation field is said to be black body radiation, and is described by Planck's law. For local thermodynamic equilibrium, the radiation field does not have be a black body field, but the rate of interatomic collisions must vastly exceed the rates of absorption and emission of quanta of light, so that the interatomic collisions entirely dominate the distribution of states of atomic excitation.

In the cases of thermodynamic equilibrium and of local thermodynamic equilibrium, the densities of the atoms, both excited and unexcited, may be calculated from the Maxwell–Boltzmann distribution, but for other cases, (e.g. lasers) the calculation is more complicated.

The Einstein Coefficients

In 1916, Albert Einstein proposed that there are essentially three processes occurring in the formation of an atomic spectral line. The three processes are referred to as spontaneous emission, stimulated emission and absorption and with each is associated an Einstein coefficient which is a measure of the probability of that particular process occurring.

Spontaneous Emission

Spontaneous emission is the process by which an electron "spontaneously" (i.e. without any outside influence) decays from a higher energy level to a lower one. The process is described by the Einstein coefficient A_{21} (s^{-1}) which gives the probability per unit time that an electron in state 2 with energy E_2 will decay spontaneously to state 1 with energy E_1, emitting a photon with an energy $E_2 - E_1 = h\nu$.

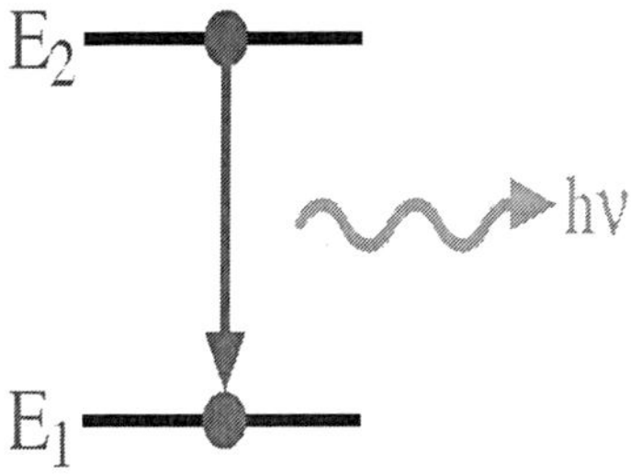

Figure 27: Schematic diagram of atomic spontaneous emission

Due to the energy-time uncertainty principle, the transition actually produces photons within a narrow range of frequencies called the spectral linewidth. If n_i is the number density of atoms in state i then the change in the number density of atoms in state 1 per unit time due to spontaneous emission will be:

$$\left(\frac{dn_1}{dt}\right)_{A_{21}} = A_{21} n_2$$

Stimulated Emission

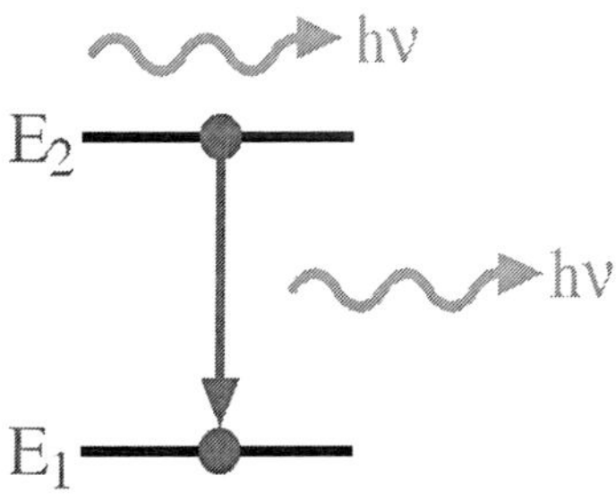

Figure 28: Schematic diagram of atomic stimulated emission

Stimulated emission (also known as induced emission) is the process by which an electron is induced to jump from a higher energy level to a lower one by the presence of electromagnetic radiation at (or near) the frequency of the transition. The process is described by the Einstein coefficient B_{21} ($\mathrm{sr \cdot m^2 \cdot Hz \cdot W^{-1} \cdot s^{-1} = sr \cdot m^2 \cdot J^{-1} \cdot s^{-1}}$), which gives the probability per unit time per unit spectral radiance of the radiation field that an electron in state 2 with energy E_2 will decay to state 1 with energy E_1, emitting a photon with an energy $E_2 - E_1 = h\nu$. The change in the number density of atoms in state 1 per unit time due to induced emission will be:

$$\left(\frac{dn_1}{dt}\right)_{B_{21}} = B_{21}n_2\rho(\nu); \qquad \rho(\nu) = \frac{2h\nu^3}{c^2(e^{h\nu/kT}-1)}$$

where **r**(ν) is the radiation density of the radiation field at the frequency of the transition.

Stimulated emission is one of the fundamental processes that led to the development of the laser.

Photo Absorption

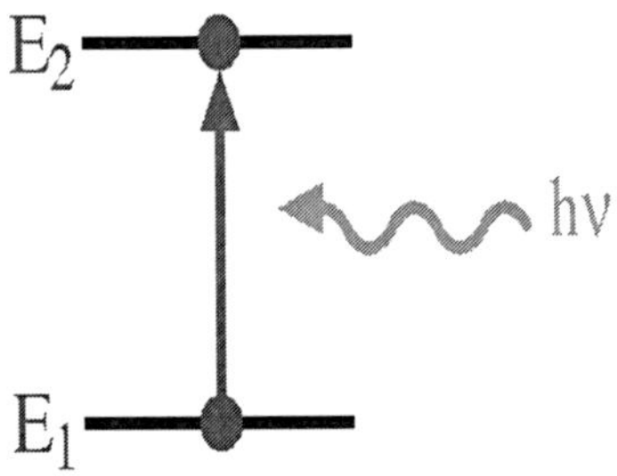

Figure 29: Schematic diagram of atomic absorption

Absorption is the process by which a photon is absorbed by the atom, causing an electron to jump from a lower energy level to a higher one. The process is described by the Einstein coefficient B_{12}(sr·m^2·Hz·W^{-1}·s^{-1} = sr·m^2·J^{-1}·s^{-1}), which gives the probability per unit time per unit spectral radiance of the radiation field that an electron in state 1 with energy E_1 will absorb a photon with an energy $E_2 - E_1 = h\nu$ and jump to state 2 with energy E_2. The change in the number density of atoms in state 1 per unit time due to absorption will be:

$$\left(\frac{dn_1}{dt}\right)_{B_{12}} = -B_{12}n_1\rho(\nu)$$

Detailed Balancing

The Einstein coefficients are fixed probabilities associated with each atom, and do not depend on the state of the gas of which the atoms are a part. Therefore, any relationship that we can derive between the coefficients at, say, thermal equilibrium will be valid universally.

At equilibrium, we will have a simple balancing, in which the net change in the number of any excited atoms is zero, being balanced by loss and gain due to all processes. With respect to bound-bound transitions, we will have detailed balancing as well, which states that the net exchange between any two levels will be balanced. This is because the probabilities of transition cannot be affected by the presence or absence of other excited atoms. Detailed balance (valid only at equilibrium) requires that the change in time of the number of atoms in level 1 due to the above three processes be zero:

$$0 = A_{21}n_2 + B_{21}n_2 I(\nu) - B_{12}n_1 I(\nu)$$

Along with detailed balancing, we may use our knowledge of the equilibrium energy distribution of the atoms, as stated in the Maxwell–Boltzmann distribution , and the equilibrium distribution of the photons, as stated in Planck's law of black body radiation to derive universal relationships between the Einstein coefficients.

From the Maxwell–Boltzmann distribution we have for the number of excited atomic specie i:

$$\frac{n_i}{n} = \frac{g_i e^{-E_i/kT}}{Z}$$

where n is the total density of the atomic specie, excited and unexcited, k is Boltzmann's constant, T is the temperature, g_i is the degeneracy of state i, and Z is the partition function. From Planck's law of black body radiation we have for thespectral radiance at frequency ν

$$I(\nu) = \frac{F(\nu)}{e^{h\nu/kT} - 1}$$

where:

$$F(\nu) = \frac{2h\nu^3}{c^2}$$

where c is the speed of light and h is Planck's constant. Note that in some treatments, the blackbody energy density is used rather than the spectral radiance, in which case:

$$F(\nu) = \frac{8\pi h\nu^3}{c^3}$$

Substituting these expressions into the equation of detailed balancing and remembering that $E_2 — E_1 = h\nu$ yields:

$$A_{21}g_2 e^{-h\nu/kT} + B_{21}g_2 e^{-h\nu/kT} \frac{F(\nu)}{e^{h\nu/kT} - 1} = B_{12}g_1 \frac{F(\nu)}{e^{h\nu/kT} - 1}$$

separating to:

$$A_{21}g_2(e^{h\nu/kT} - 1) + B_{21}g_2 F(\nu) = B_{12}g_1 e^{h\nu/kT} F(\nu)$$

The above equation must hold at any temperature, so

$$A_{21}g_2 = B_{12}g_1F(\nu)$$
$$-A_{21}g_2 + B_{21}g_2F(\nu) = 0$$

Therefore the three Einstein coefficients are interrelated by:

$$\frac{A_{21}}{B_{21}} = F(\nu)$$

and

$$\frac{B_{21}}{B_{12}} = \frac{g_1}{g_2}$$

When this relation is inserted into the original equation, one can also find a relation between A_{21} and B_{12}, involving Planck's law.

Oscillator Strengths

The oscillator strength f_{12} *is defined by the following relation to the cross section* a_{12} *for absorption:*

$$a_{12} = \frac{\pi e^2}{m_e c} f_{12}$$

where e is the electron charge and m_e is the electron mass. This allows all three Einstein coefficients to be expressed in terms of the single oscillator strength associated with the particular atomic spectral line:

$$B_{12} = \frac{4\pi^2 e^2}{m_e h\nu c} f_{12}$$

$$B_{21} = \frac{4\pi^2 e^2}{m_e h\nu c} \frac{g_1}{g_2} f_{12}$$

$$A_{21} = \frac{8\nu^2 \pi^2 e^2}{m_e c^3} \frac{g_1}{g_2} f_{12}$$

Transmission Electron Microscopy

Transmission electron microscopy (TEM) is a microscopy technique whereby a beam of electrons is transmitted through an ultra thin specimen, interacting with the specimen as it passes through. An image is formed from the interaction of the electrons transmitted through the specimen; the image is magnified and focused onto an imaging device, such as a fluorescent screen, on a layer of photographic film, or to be detected by a sensor such as a CCD camera. TEMs are capable of imaging at a significantly higher resolution than light microscopes, owing to the small de Broglie wavelength of electrons. This enables the instrument's user to examine fine detail—even as small as a single column of atoms, which is tens of thousands times smaller than the smallest resolvable object in a light microscope.

TEM forms a major analysis method in a range of scientific fields, in both physical and biological sciences. TEMs find application in cancer research, virology, materials science as well as pollution, nanotechnology, and semiconductor research.

At smaller magnifications TEM image contrast is due to absorption of electrons in the material, due to the thickness and composition of the material. At higher magnifications complex wave interactions modulate the intensity of the image, requiring expert analysis of observed images. Alternate modes of use allow for the TEM to observe modulations in chemical identity, crystal orientation, electronic structure and sample induced electron phase shift as well as the regular absorption based imaging.

The first TEM was built by Max Knoll and Ernst Ruska in 1931, with this group developing the first TEM with resolving power greater than that of light in 1933 and the first commercial TEM in 1939.

Initial Development

Figure 30: The first practical TEM, Originally installed at I. G Farben-Werke and now on display at the Deutsches Museum in Munich, Germany

Ernst Abbe originally proposed that the ability to resolve detail in an object was limited by the wavelength of the light used in imaging, thus limiting the useful obtainable magnification from an optical microscope to a few micrometers. Developments into ultraviolet (UV) microscopes, led by Köhler and Rohr, allowed for an increase in resolving power of about a factor of two. However this required more expensive quartz optical components, due to the absorption of UV by glass. At this point it was believed that obtaining an image with sub-micrometer information was simply impossible due to this wavelength constraint.

It had earlier been recognised by Plücker in 1858 that the deflection of "cathode rays" (electrons) was possible by the use of magnetic fields. This effect had been utilised to build primitive cathode ray oscilloscopes (CROs) as early as 1897 by Ferdinand Braun, intended as a measurement device. Indeed in 1891 it was recognised by Riecke that the cathode rays could be focused by these magnetic fields, allowing for simple lens designs. Later this theory was extended by Hans Busch in his work published in 1926, who showed that the lens maker's equation, could under appropriate assumptions, be applicable to electrons.

In 1928, at the Technological University of Berlin Adolf Matthias, Professor of High voltage Technology and Electrical Installations, appointed Max Knoll to lead a team of researchers to advance the CRO design. The team consisted of several PhD students including Ernst Ruska and Bodo von Borries. This team of researchers concerned themselves with lens design and CRO column placement, which they attempted to obtain the parameters that could be optimised to allow for construction of better CROs, as well as the development of electron optical components which could be used to generate low magnification (nearly 1:1) images. In 1931 the group successfully generated magnified images of mesh grids placed over the anode aperture. The device used two magnetic lenses to achieve higher magnifications, arguably the first electron microscope. In that same year, Reinhold Rudenberg, the scientific director of the Siemens company, had patented an electrostatic lens electron microscope.

Improving Resolution

At this time the wave nature of electrons, which were considered charged matter particles, had not been fully realised until the publication of the De Broglie hypothesis in 1927. The group was unaware of this publication until 1932, where it was quickly realised that the De Broglie wavelength of electrons was many orders of magnitude smaller than that for light, theoretically allowing for imaging at atomic scales. In April 1932, Ruska suggested the construction of a new electron microscope for direct imaging of specimens inserted into the microscope, rather than simple mesh grids or images of apertures. With this device successful diffraction and normal imaging of aluminium sheet was achieved, however exceeding the magnification achievable with light microscopy had still not been successfully demonstrated. This goal was achieved in September 1933, using images of cotton fibers, which were quickly acquired before being damaged by the electron beam.

At this time, interest in the electron microscope had increased, with other groups, such as Albert Prebus and James Hillier at the University of Toronto who constructed the first TEM in North America in 1938, continually advancing TEM design.

Research continued on the electron microscope at Siemens in 1936, the aim of the research was the development improvement of TEM imaging properties, particularly with regard to biological specimens. At this time electron microscopes were being fabricated for specific groups, such as the "EM1" device used at the UK National Physical Laboratory. In 1939 the first commercial electron microscope, pictured, was installed in the Physics department of I. G Farben-Werke. Further work on the electron microscope was hampered by the destruction of a new laboratory constructed at Siemens by an air-raid, as well as the death of two of the researchers, Heinz Müller and Friedrick Krause during World War II.

Further Research

After World War II, Ruska resumed work at Siemens, where he continued to develop the electron microscope, producing the first microscope with 100k magnification. The fundamental structure of this microscope design, with multi-stage beam preparation optics, is still used in modern microscopes. The worldwide electron microscopy community advanced with electron microscopes being manufactured in Manchester UK, the USA (RCA), Germany (Siemens) and Japan . The first international conference in electron microscopy was in Delft in 1942, with more than one hundred attendees. Later conferences included the "First" international conference in Paris, 1950 and then in London in 1954.

With the development of TEM, the associated technique of scanning transmission electron microscopy (STEM) was re-investigated and did not become developed until the 1970s, with Albert Crewe at the University of Chicago developing the field emission gun and adding a high quality objective lens to create the modern STEM. Using this design, Crewe demonstrated the ability to image atoms using annular dark-field imaging. Crewe and coworkers at the University of Chicago developed the cold field electron emission source and built a STEM able to visualise single heavy atoms on thin carbon substrates.

Background

Electrons

Theoretically, the maximum resolution, d, that one can obtain with a light microscope has been limited by the wavelength of the photons that are being used to probe the sample, λ and the numerical aperture of the system, *NA*.

$$d = \frac{\lambda}{2n\sin\alpha} \approx \frac{\lambda}{2\mathrm{NA}}$$

Early twentieth century scientists theorised ways of getting around the limitations of the relatively large wavelength of visible light (wavelengths of 400–700 nanometers) by using electrons. Like all matter, electrons have both wave and particle properties (as theorised by Louis-Victor de Broglie), and their wavelike properties mean that a beam of electrons can be made to behave like a beam of electromagnetic radiation. The wavelength of electrons is found by equating the de Broglie equation to the kinetic energy of an electron. An additional correction must be made to account for relativistic effects, as in a TEM an electron's velocity approaches the speed of light, c.

$$\lambda_e \approx \frac{h}{\sqrt{2m_0E\left(1 + \frac{E}{2m_0c^2}\right)}}$$

where, h is Planck's constant, m_0 is the rest mass of an electron and E is the energy of the accelerated electron. Electrons are usually generated in an electron microscope by a process known as thermionic emission from a filament, usually tungsten, in the same manner as a light bulb, or alternatively by field electron emission. The electrons are then accelerated by an electric potential (measured in volts) and focused by electrostatic and electromagnetic lenses onto the sample.

The transmitted beam contains information about electron density, phase and periodicity; this beam is used to form an image.

Source Formation

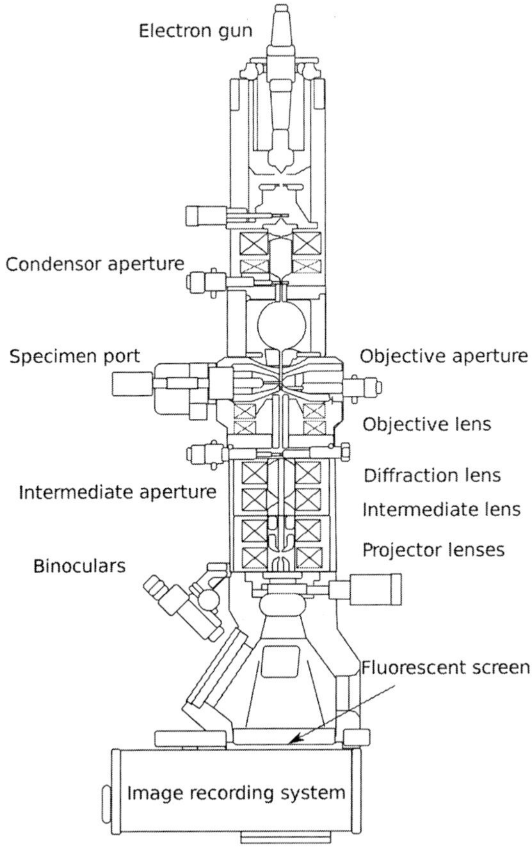

Figure 31: Layout of optical components in a basic TEM

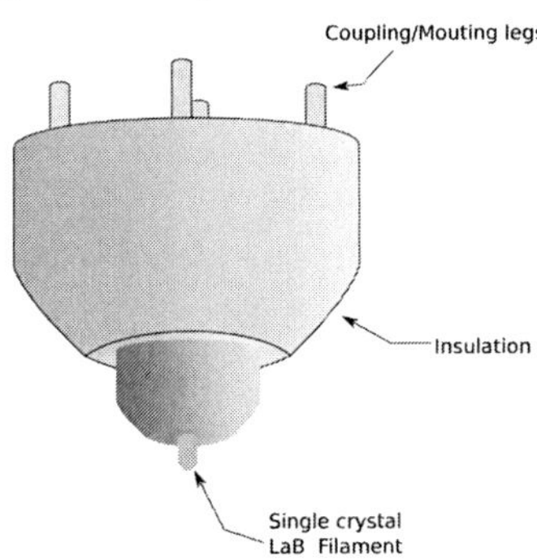

Figure 32: Single crystal LaB_6 filament

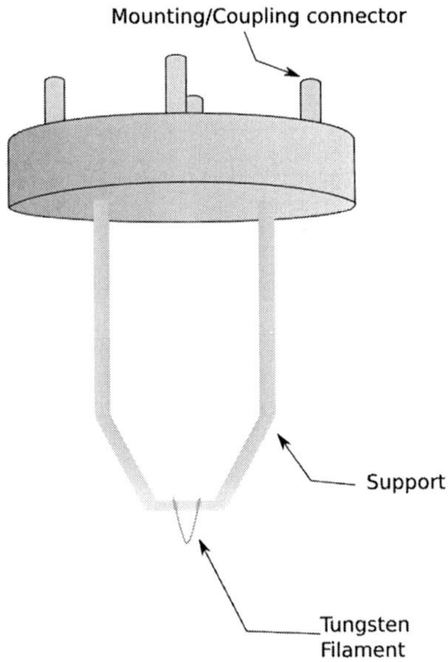

Figure 33: Hairpin style tungsten filament

From the top down, the TEM consists of an emission source, which may be a tungsten filament, or a lanthanum hexaboride (LaB_6) source. For tungsten, this will be of the form of either a hairpin-style filament, or a small spike-shaped filament. LaB_6 sources utilise small single crystals. By connecting this gun to a high voltage source (typically ~100-300 kV) the gun will, given sufficient current, begin to emit electrons either by thermionic or field electron emission into the vacuum. This extraction is usually aided by the use of a Wehnelt cylinder. Once extracted, the upper lenses of the TEM allow for the formation of the electron probe to the desired size and location for later interaction with the sample.

Manipulation of the electron beam is performed using two physical effects. The interaction of electrons with a magnetic field will cause electrons to move according to the right hand rule, thus allowing for electromagnets to manipulate the electron beam. The use of magnetic fields allows for the formation of a magnetic lens of variable focusing power, the lens shape originating due to the distribution of magnetic flux. Additionally, electrostatic fields can cause the electrons to be deflected through a constant angle. Coupling of two deflections in opposing directions with a small intermediate gap allows for the formation of a shift in the beam path, this being used in TEM for beam shifting, subsequently this is extremely important to STEM. From these two effects, as well as the use of an electron imaging system, sufficient control over the beam path is possible for TEM operation. The optical configuration of a TEM can be rapidly changed, unlike that for an optical microscope, as lenses in the beam path can be enabled, have their strength changed, or be disabled entirely simply via rapid electrical switching, the speed of which is limited by effects such as the magnetic hysteresis of the lenses.

Optics

The lenses of a TEM allow for beam convergence, with the angle of convergence as a variable parameter, giving the TEM the ability to change magnification simply by modifying the amount of current that flows through the coil, quadrupole or hexapole lenses. The quadrupole lens is an arrangement of electromagnetic coils at the vertices of the square, enabling the generation of a lensing magnetic fields, the hexapole configuration simply enhances the lens symmetry by using six, rather than four coils.

Typically a TEM consists of three stages of lensing. The stages are the condensor lenses, the objective lenses, and the projector lenses. The condensor lenses are responsible for primary beam formation, whilst the objective lenses focus the beam that comes through the sample itself (in STEM scanning mode, there are also objective lenses above the sample to make the incident electron beam convergent). The projector lenses are used to expand the beam onto the phosphor screen or other imaging device, such as film. The magnification of the TEM is due to the ratio of the distances between the specimen and the objective lens' image plane. Additional quad or hexapole lenses allow for the correction of asymmetrical beam distortions, known as astigmatism. It is noted that TEM optical configurations differ significantly with implementation, with manufacturers using custom lens configurations, such as in spherical aberration corrected instruments, or TEMs utilising energy filtering to correct electron chromatic aberration.

Display

Imaging systems in a TEM consist of a phosphor screen, which may be made of fine (10-100 ìm) particulate zinc sulphide, for direct observation by the operator. Optionally, an image recording system such as film based or doped YAGscreen coupled CCDs. Typically these devices can be removed or inserted into the beam path by the operator as required.

Components

Figure 34: The electron source of the TEM is at the top, where the lensing system (4,7 and 8) focuses the beam on the specimen and then projects it onto the viewing screen (10). The beam control is on the right (13 and 14)

A TEM is composed of several components, which include a vacuum system in which the electrons travel, an electron emission source for generation of the electron stream, a series of electromagnetic lenses, as well as electrostatic plates. The latter two allow the operator to guide and manipulate the beam as required. Also required is a device to allow the insertion into, motion within, and removal of specimens from the beam path. Imaging devices are subsequently used to create an image from the electrons that exit the system.

Vacuum System

To increase the mean free path of the electron gas interaction, a standard TEM is evacuated to low pressures, typically on the order of 10^{-4} Pa. The need for this is twofold: first the allowance for the voltage difference between the cathode and the ground without generating an arc, and secondly to reduce the collision frequency of electrons with gas atoms to negligible levels—this effect is characterised by the mean free path. TEM components such as specimen holders and film cartridges must be routinely inserted or replaced requiring a system with the ability to re-evacuate on a regular basis. As such,

TEMs are equipped with multiple pumping systems and airlocks and are not permanently vacuum sealed. The vacuum system for evacuating a TEM to an operating pressure level consists of several stages. Initially a low or roughing vacuum is achieved with either a rotary vane pump or diaphragm pumpsbringing the TEM to a sufficiently low pressure to allow the operation of a turbomolecular or diffusion pump which brings the TEM to its high vacuum level necessary for operations. To allow for the low vacuum pump to not require continuous operation, while continually operating the turbomolecular pumps, the vacuum side of a low-pressure pump may be connected to chambers which accommodate the exhaust gases from the turbomolecular pump. Sections of the TEM may be isolated by the use of gate valves, to allow for different vacuum levels in specific areas, such as a higher vacuum of 10^{-4} to 10^{-7} Pa or higher in the electron gun in high-resolution or field-emission TEMs.

High-voltage TEMs require ultra-high vacuums on the range of 10^{-7} to 10^{-9} Pa to prevent generation of an electrical arc, particularly at the TEM cathode. As such for higher voltage TEMs a third vacuum system may operate, with the gun isolated from the main chamber either by use of gate valves or by the use of a differential pumping aperture. The differential pumping aperture is a small hole that prevents diffusion of gas molecules into the higher vacuum gun area faster than they can be pumped out. For these very low pressures either an ion pump or a getter material is used. Poor vacuum in a TEM can cause several problems, from deposition of gas inside the TEM onto the specimen as it is being viewed through a process known as electron beam induced deposition, or in more severe cases damage to the cathode from an electrical discharge. Vacuum problems due to specimen sublimation are limited by the use of a cold trap to adsorb sublimated gases in the vicinity of the specimen.

Specimen Stage

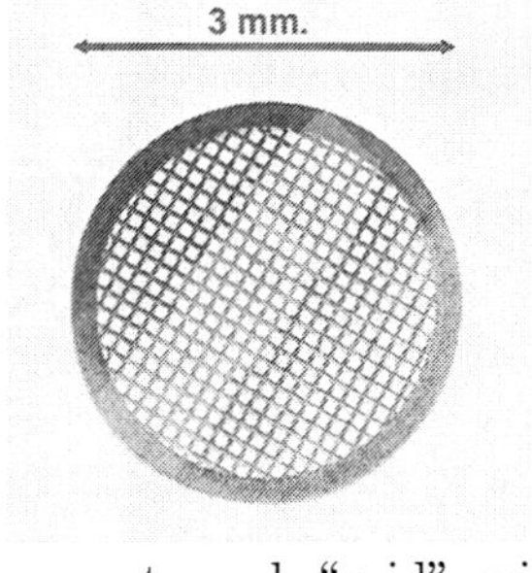

Figure 35: TEM sample support mesh "grid", with ultramicrotomy sections

TEM specimen stage designs include airlocks to allow for insertion of the specimen holder into the vacuum with minimal increase in pressure in other areas of the microscope. The specimen holders are adapted to hold a standard size of grid upon which the sample is placed or a standard size of self-supporting specimen. Standard TEM grid sizes is a 3.05 mm diameter ring, with a thickness and mesh size ranging from a few to 100 μm. The sample is placed onto the inner meshed area having diameter of approximately 2.5 mm. Usual grid materials are copper, molybdenum, gold or platinum. This grid is placed into the sample holder which is paired with the specimen stage. A wide variety of designs of stages and holders exist, depending upon the type of experiment being performed. In addition to 3.05 mm grids, 2.3 mm grids are sometimes, if rarely, used. These grids were particularly used in the mineral sciences where a large degree of tilt can be required and where specimen material may be extremely rare. Electron transparent specimens have a thickness around 100 nm, but this value depends on the accelerating voltage.

Once inserted into a TEM, the sample often has to be manipulated to present the region of interest to the beam, such as in single grain diffraction, in a specific orientation. To accommodate this, the TEM stage includes mechanisms for the translation of the sample in the XY plane of the sample, for Z height adjustment of the sample holder, and usually for at least one rotation degree of freedom for the sample. Thus a TEM stage may provide four degrees of freedom for the motion of the specimen. Most modern TEMs provide the ability for two orthogonal rotation angles of movement with specialised holder designs called double-tilt sample holders. Of note however is that some stage designs, such as top-entry or vertical insertion stages once common for high resolution TEM studies, may simply only have X-Y translation available. The design criteria of TEM stages are complex, owing to the simultaneous requirements of mechanical and electron-optical constraints and have thus generated many unique implementations.

A TEM stage is required to have the ability to hold a specimen and be manipulated to bring the region of interest into the path of the electron beam. As the TEM can operate over a wide range of magnifications, the stage must simultaneously be highly resistant to mechanical drift, with drift requirements as low as a few nm/minute while being able to move several um/minute, with repositioning accuracy on the order of nanometers. Earlier designs of TEM accomplished this with a complex set of mechanical downgearing devices, allowing the operator to finely control the motion of the stage by several rotating rods. Modern devices may use electrical stage designs, using screw gearing in concert with stepper motors, providing the operator with a computer-based stage input, such as a joystick or trackball.

Two main designs for stages in a TEM exist, the side-entry and top entry version. Each design must accommodate the matching holder to allow for specimen insertion without either damaging delicate TEM optics or allowing gas into TEM systems under vacuum.

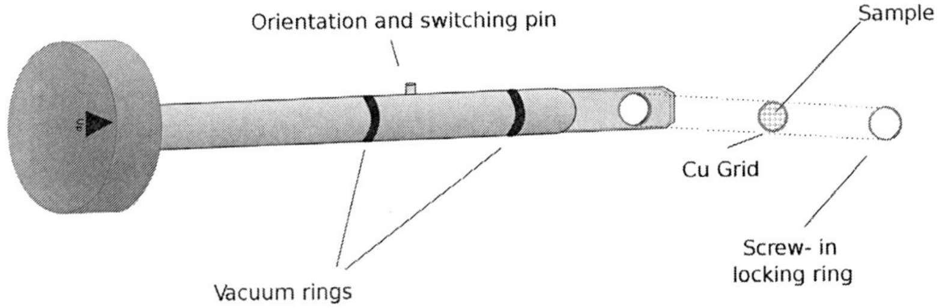

Figure 36: A diagram of a single axis tilt sample holder for insertion into a TEM goniometer. Titling of the holder is achieved by rotation of the entire goniometer

The most common is the side entry holder, where the specimen is placed near the tip of a long metal (brass or stainless steel) rod, with the specimen placed flat in a small bore. Along the rod are several polymer vacuum rings to allow for the formation of a vacuum seal of sufficient quality, when inserted into the stage. The stage is thus designed to accommodate the rod, placing the sample either in between or near the objective lens, dependent upon the objective design. When inserted into the stage, the side entry holder has its tip contained within the TEM vacuum, and the base is presented to atmosphere, the airlock formed by the vacuum rings.

Insertion procedures for side entry TEM holders typically involve the rotation of the sample to trigger micro switches that initiate evacuation of the airlock before the sample is inserted into the TEM column.

The second design is the top-entry holder consists of a cartridge that is several cm long with a bore drilled down the cartridge axis. The specimen is loaded into the bore, possibly utilising a small screw ring to hold the sample in place. This cartridge is inserted into an airlock with the bore perpendicular to the TEM optic axis. When sealed, the airlock is manipulated to push the cartridge such that the cartridge falls into place, where the bore hole becomes aligned with the beam axis, such that the beam travels down the cartridge bore and into the specimen. Such designs are typically unable to be tilted without blocking the beam path or interfering with the objective lens.

Electron Gun

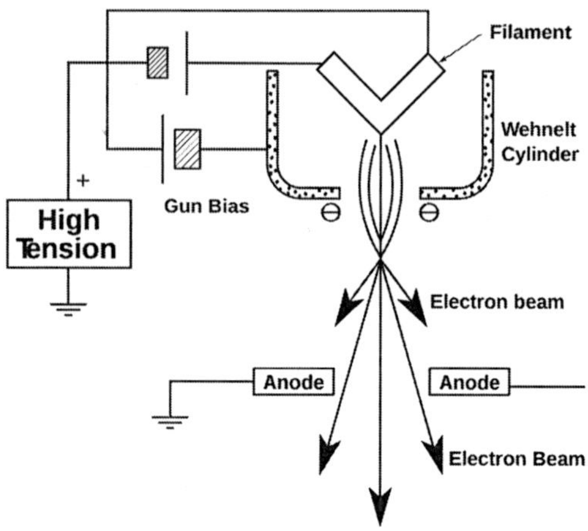

Figure 37: Cross sectional diagram of an electron gun assembly, illustrating electron extraction

The electron gun is formed from several components: the filament, a biasing circuit, a Wehnelt cap, and an extraction anode. By connecting the filament to the negative component power supply, electrons can be "pumped" from the electron gun to the anode plate, and TEM column, thus completing the circuit. The gun is designed to create a beam of electrons exiting from the assembly at some given angle, known as the gun divergence semiangle, **a**. By constructing the Wehnelt cylinder such that it has a higher negative charge than the filament itself, electrons that exit the filament in a diverging manner are, under proper operation, forced into a converging pattern the minimum size of which is the gun crossover diameter.

The thermionic emission current density, J, can be related to the work function of the emitting material and is a Boltzmann distribution given below, where A is a constant, Ö is the work function and T is the temperature of the material.

$$J = AT^2 \exp\left(-\frac{\Phi}{kT}\right)$$

This equation shows that in order to achieve sufficient current density it is necessary to heat the emitter, taking care not to cause damage by application of excessive heat, for this reason materials with either a high melting point, such as tungsten, or those with a low work function (LaB_6) are required for the gun filament. Furthermore both lanthanum hexaboride and tungsten thermionic sources must be heated in order to achieve thermionic emission, this can be achieved by the use of a small resistive strip. To prevent thermal shock, there is often a delay enforced in the application of current to the tip, to prevent thermal gradients from damaging the filament, the delay is usually a few seconds for LaB_6, and significantly lower for tungsten.

Electron Lens

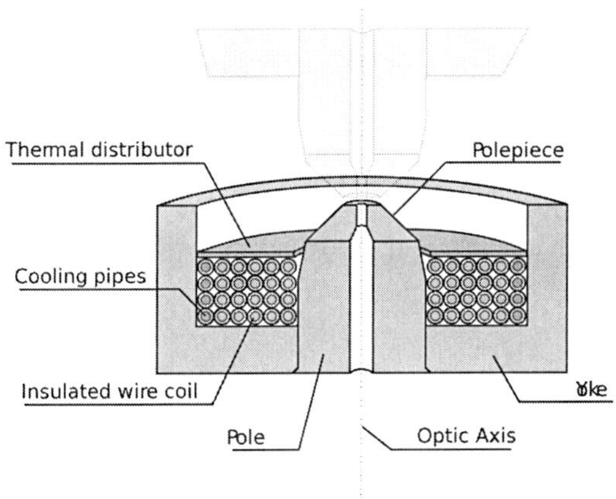

Figure 38: Diagram of a TEM split polepiece design lens

Electron lenses are designed to act in a manner emulating that of an optical lens, by focusing parallel rays at some constant focal length. Lenses may operate electrostatically or magnetically. The majority of electron lenses for TEM utilise electromagnetic coils to generate a convex lens. For these lenses the field produced for the lens must be radially symmetric, as deviation from the radial symmetry of the magnetic lens causes aberrations such as astigmatism, and worsens spherical and chromatic aberration. Electron lenses are manufactured from iron, iron-cobalt or nickel cobalt alloys, such as permalloy. These are selected for their magnetic properties, such as magnetic saturation, hysteresis and permeability.

The components include the yoke, the magnetic coil, the poles, the polepiece, and the external control circuitry. The polepiece must be manufactured in a very symmetrical manner, as this provides the boundary conditions for the magnetic field that forms the lens. Imperfections in the manufacture of the polepiece can induce severe distortions in the magnetic field symmetry, which induce distortions that will ultimately limit the lenses' ability to reproduce the object plane. The exact dimensions of the gap, pole piece internal diameter and taper, as well as the overall design of the lens is often performed by finite element analysis of the magnetic field, whilst considering the thermal and electrical constraints of the design.

The coils which produce the magnetic field are located within the lens yoke. The coils can contain a variable current, but typically utilise high voltages, and therefore require significant insulation in order to prevent short-circuiting the lens components. Thermal distributors are placed to ensure the extraction of the heat generated by the energy lost to resistance of the coil windings. The windings may be water-cooled, using a chilled water supply in order to facilitate the removal of the high thermal duty.

Apertures

Apertures are annular metallic plates, through which electrons that are further than a fixed distance from the optic axis may be excluded. These consist of a small metallic disc that is sufficiently thick to prevent electrons from passing through the disc, whilst permitting axial electrons. This permission of central electrons in a TEM causes two effects simultaneously: firstly, apertures decrease the beam intensity as electrons are filtered from the beam, which may be desired in the case of beam sensitive samples. Secondly, this filtering removes electrons that are scattered to high angles, which may be due to unwanted processes such as spherical or chromatic aberration, or due to diffraction from interaction within the sample.

Apertures are either a fixed aperture within the column, such as at the condensor lens, or are a movable aperture, which can be inserted or withdrawn from the beam path, or moved in the plane perpendicular to the beam path. Aperture assemblies are mechanical devices which allow for the selection of different aperture sizes, which may be used by the operator to trade off intensity and the filtering effect of the aperture. Aperture assemblies are often equipped with micrometers to move the aperture, required during optical calibration.

Imaging Methods

Imaging methods in TEM utilise the information contained in the electron waves exiting from the sample to form an image. The projector lenses allow for the correct positioning of this electron wave distribution onto the viewing system. The observed intensity of the image, I, assuming sufficiently high quality of imaging device, can be approximated as proportional to the time-average amplitude of the electron wavefunctions, where the wave which form the exit beam is denoted by Ø.

$$I(x) = \frac{k}{t_1 - t_0} \int_{t_0}^{t_1} \Psi\Psi^* \, dt$$

Different imaging methods therefore attempt to modify the electron waves exiting the sample in a form that is useful to obtain information with regards to the sample, or beam itself. From the previous equation, it can be deduced that the observed image depends not only on the amplitude of beam, but also on the phase of the electrons, although phase effects may often be ignored at lower magnifications. Higher resolution imaging requires thinner samples and higher energies of incident electrons. Therefore the sample can no longer be considered to be absorbing electrons, via a Beer's law effect, rather the sample can be modelled as an object that does not change the amplitude of the incoming electron wavefunction. Rather the sample modifies the phase of the incoming wave; this model is known as a pure phase object, for sufficiently thin specimens phase effects dominate the image, complicating analysis of the observed intensities. For example, to improve the contrast in the image the TEM may be operated at a slight defocus to enhance contrast, owing to convolution by the contrast transfer function of the TEM, which would normally decrease contrast if the sample was not a weak phase object.

Contrast Formation

Contrast formation in the TEM depends greatly on the mode of operation. Complex imaging techniques, which utilise the unique ability to change lens strength or to deactivate a lens, allow for many operating modes. These modes may be used to discern information that is of particular interest to the investigator.

Bright Field

The most common mode of operation for a TEM is the bright field imaging mode. In this mode the contrast formation, when considered classically, is formed directly by occlusion and absorption of electrons in the sample. Thicker regions of the sample, or regions with a higher atomic number will appear dark, whilst regions with no sample in the beam path will appear bright – hence the term "bright field". The image is in effect assumed to be a simple two dimensional projection of the sample down the optic axis, and to a first approximation may be modelled via Beer's law, more complex analyses require the modelling of the sample to include phase information.

Diffraction Contrast

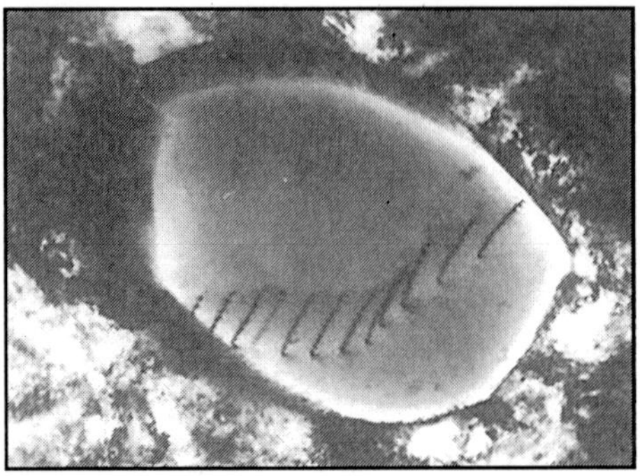

Figure 39: Transmission electron micrograph of dislocations in steel, which are faults in the structure of the crystal lattice at the atomic scale

Samples can exhibit diffraction contrast, whereby the electron beam undergoes Bragg scattering, which in the case of a crystalline sample, disperses electrons into discrete locations in the back focal plane. By the placement of apertures in the back focal plane, i.e. the objective aperture, the desired Bragg reflections can be selected (or excluded), thus only parts of the sample that are causing the electrons to scatter to the selected reflections will end up projected onto the imaging apparatus.

If the reflections that are selected do not include the unscattered beam (which will appear up at the focal point of the lens), then the image will appear dark wherever no sample scattering to the selected peak is present, as such a region without a specimen will appear dark. This is known as a dark-field image.

Modern TEMs are often equipped with specimen holders that allow the user to tilt the specimen to a range of angles in order to obtain specific diffraction conditions, and apertures placed above the specimen allow the user to select electrons that would otherwise be diffracted in a particular direction from entering the specimen.

Applications for this method include the identification of lattice defects in crystals. By carefully selecting the orientation of the sample, it is possible not just to determine the position of defects but also to determine the type of defect present. If the sample is oriented so that one particular plane is only slightly tilted away from the strongest diffracting angle (known as the Bragg Angle), any distortion of the crystal plane that locally tilts the plane to the Bragg angle will produce particularly strong contrast variations. However, defects that produce only displacement of atoms that do not tilt the crystal to the Bragg angle (i. e. displacements parallel to the crystal plane) will not produce strong contrast.

Electron Energy Loss

Utilising the advanced technique of EELS, for TEMs appropriately equipped electrons can be rejected based upon their voltage (which, due to constant charge is their energy), using magnetic sector based devices known as EELS spectrometers. These devices allow for the selection of particular energy values, which can be associated with the way the electron has interacted with the sample. For example different elements in a sample result in different electron energies in the beam after the sample. This normally results in chromatic aberration – however this effect can, for example, be used to generate an image which provides information on elemental composition, based upon the atomic transition during electron-electron interaction.

EELS spectrometers can often be operated in both spectroscopic and imaging modes, allowing for isolation or rejection of elastically scattered beams. As for many images inelastic scattering will include information that may not be of interest to the investigator thus reducing observable signals of interest, EELS imaging can be used to enhance contrast in observed images, including both bright field and diffraction, by rejecting unwanted components.

Phase Contrast

Crystal structure can also be investigated by high-resolution transmission electron microscopy (HRTEM), also known as phase contrast. When utilising a Field emission source, of uniform thickness, the images are formed due to differences in phase of electron waves, which is caused by specimen interaction. Image formation is given by the complex modulus of the incoming electron beams. As such, the image is not only dependent on the number of electrons hitting the screen, making direct interpretation of phase contrast images more complex. However this effect can be used to an advantage, as it can be manipulated to provide more information about the sample, such as in complex phase retrieval techniques.

Diffraction

As previously stated, by adjusting the magnetic lenses such that the back focal plane of the lens rather than the imaging plane is placed on the imaging apparatus a diffraction pattern can be generated. For thin crystalline samples, this produces an image that consists of a pattern of dots in the case of a single crystal, or a series of rings in the case of a polycrystalline or amorphous solid material. For the single crystal case the diffraction pattern is dependent upon the orientation of the specimen and the structure of the sample illuminated by the electron beam. This image provides the investigator with information about the space group symmetries in the crystal and the crystal's orientation to the beam path. This is typically done without utilising any information but the position at which the diffraction spots appear and the observed image symmetries.

Diffraction patterns can have a large dynamic range, and for crystalline samples, may have intensities greater than those recordable by CCD. As such, TEMs may still be equipped with film cartridges for the purpose of obtaining these images, as the film is a single use detector.

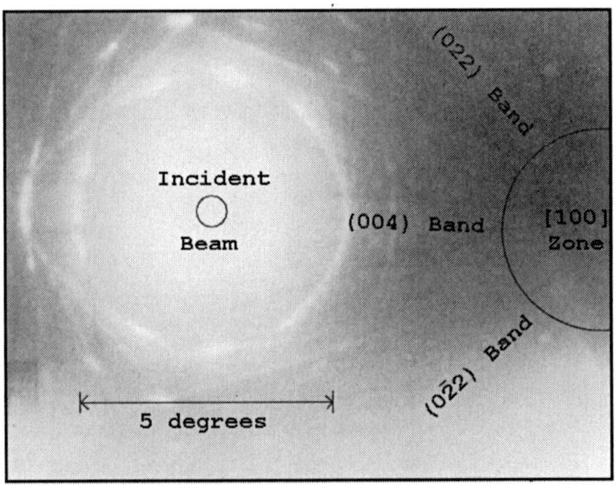

Figure 40: Convergent Beam Kikuchi lines from Silicon, near the zone axis

Analysis of diffraction patterns beyond point-position can be complex, as the image is sensitive to a number of factors such as specimen thickness and orientation, objective lens defocus, spherical and chromatic aberration. Although quantitative interpretation of the contrast shown in lattice images is possible, it is inherently complicated and can require extensive computer simulation and analysis, such aselectron multislice analysis.

More complex behaviour in the diffraction plane is also possible, with phenomena such as Kikuchi lines arising from multiple diffraction within the crystalline lattice. In convergent beam electron diffraction(CBED) where a non-parallel, i.e. converging, electron wavefront is produced by concentrating the electron beam into a fine probe at the sample surface, the interaction of the convergent beam can provide information beyond structural data such as sample thickness.

Three-dimensional Imaging

As TEM specimen holders typically allow for the rotation of a sample by a desired angle, multiple views of the same specimen can be obtained by rotating the angle of the sample along an axis perpendicular to the beam. By taking multiple images of a single TEM sample at differing angles, typically in 1° increments, a set of images known as a "tilt series" can be collected. This methodology was proposed in the 1970s by Walter Hoppe. Under purely absorption contrast conditions, this set of images can be used to construct a three-dimensional representation of the sample.

The reconstruction is accomplished by a two-step process, first images are aligned to account for errors in the positioning of a sample; such errors can occur due to vibration or mechanical drift. Alignment methods use image registration algorithms, such as autocorrelation methods to correct these errors. Secondly, using a technique known as filtered back projection, the aligned image slices can be transformed from a set of two-dimensional images, $I_j(x, y)$, to a single three-dimensional image, $I'_j(x, y, z)$. This three-dimensional image is of particular interest when morphological information is required, further study can be undertaken using computer algorithms, such as isosurfaces and data slicing to analyse the data.

As TEM samples cannot typically be viewed at a full 180° rotation, the observed images typically suffer from a "missing wedge" of data, which when using Fourier-based back projection methods decreases the range of resolvable frequencies in the three-dimensional reconstruction. Mechanical refinements, such as multi-axis tilting (two tilt series of the same specimen made at orthogonal directions) and conical tomography (where the specimen is first tilted to a given fixed angle and then imaged at equal angular rotational increments through one complete rotation in the plane of the specimen grid) can be used to limit the impact of the missing data on the observed specimen morphology. In addition, numerical techniques exist which can improve the collected data.

Non-tomographic variants on this method, referred to as single particle analysis, use images of multiple (hopefully) identical objects at different orientations to produce the image data required for three-dimensional reconstruction. If the objects do not have significant preferred orientations, this method does not suffer from the missing data wedge or cone accompanying tomographic methods, however it assumes that the different objects imaged can be treated as if the data was generated from a single object.

Sample Preparation

Sample preparation in TEM can be a complex procedure. TEM specimens are required to be at most hundreds of nanometers thick, as unlike neutron or X-Ray radiation the electron beam interacts readily with the sample, an effect that increases roughly with atomic number squared (z^2). High quality samples will have a thickness that is comparable to the mean free path of the electrons that travel through the samples, which may be only a few tens of nanometers. Preparation of TEM specimens is specific to the material under analysis and the desired information to obtain from the specimen. As such, many generic techniques have been used for the preparation of the required thin sections.

Materials that have dimensions small enough to be electron transparent, such as powders or nanotubes, can be quickly prepared by the deposition of a dilute sample containing the specimen onto support grids or films. In the biological sciences in order to withstand the instrument vacuum and facilitate handling, biological specimens can be fixated using either a negative staining material such as uranyl acetate or by plastic embedding. Alternately samples may be held at liquid nitrogen temperatures after embedding in vitreous ice. In material science and metallurgy the specimens tend to be naturally resistant to vacuum, but still must be prepared as a thin foil, or etched so some portion of the specimen is thin enough for the beam to penetrate. Constraints on the thickness of the material may be limited by the scattering cross-section of the atoms from which the material is comprised.

Tissue Sectioning

By passing samples over a glass or diamond edge, small, thin sections can be readily obtained using a semi-automated method. This method is used to obtain thin, minimally deformed samples that allow for the observation of tissue samples. Additionally inorganic samples have been studied, such as aluminium, although this usage is limited owing to the heavy damage induced in the less soft samples. To prevent charge build-up at the sample surface, tissue samples need to be coated with a thin layer of conducting material, such as carbon, where the coating thickness is several nanometers. This may be achieved via an electric arc deposition process using a sputter coating device.

Sample Staining

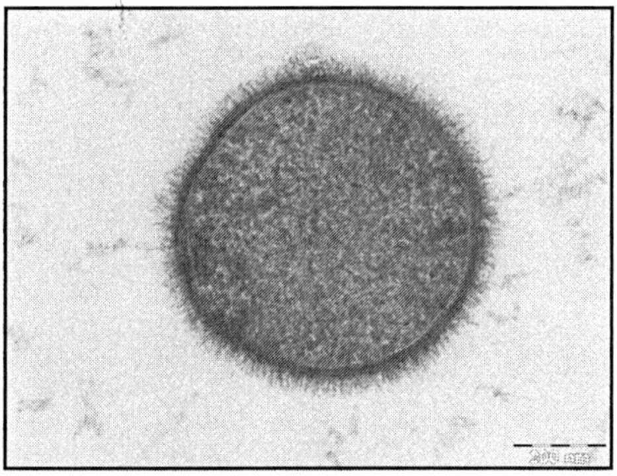

Figure 41: A section of a cell of *Bacillus subtilis*, taken with a Tecnai T-12 TEM. The scale bar is 200 nm.

Details in light microscope samples can be enhanced by stains that absorb light; similarly TEM samples of biological tissues can utilise high atomic number stains to enhance contrast. The stain absorbs electrons or scatters part of the electron beam which otherwise is projected onto the imaging system. Compounds of heavy metals such as osmium, lead, uranium or gold (in immunogold labelling) may be used prior to TEM observation to selectively deposit electron dense atoms in or on the sample in desired cellular or protein regions, requiring an understanding of how heavy metals bind to biological tissues.

Mechanical Milling

Mechanical polishing may be used to prepare samples. Polishing needs to be done to a high quality, to ensure constant sample thickness across the region of interest. A diamond, or cubic boron nitride polishing compound may be used in the final stages of polishing to remove any scratches that may cause contrast fluctuations due to varying sample thickness. Even after careful mechanical milling, additional fine methods such as ion etching may be required to perform final stage thinning.

Chemical Etching

Certain samples may be prepared by chemical etching, particularly metallic specimens. These samples are thinned using a chemical etchant, such as an acid, to prepare the sample for TEM observation. Devices to control the thinning process may allow the operator to control either the voltage or current passing through the specimen, and may include systems to detect when the sample has been thinned to a sufficient level of optical transparency.

Lon Etching

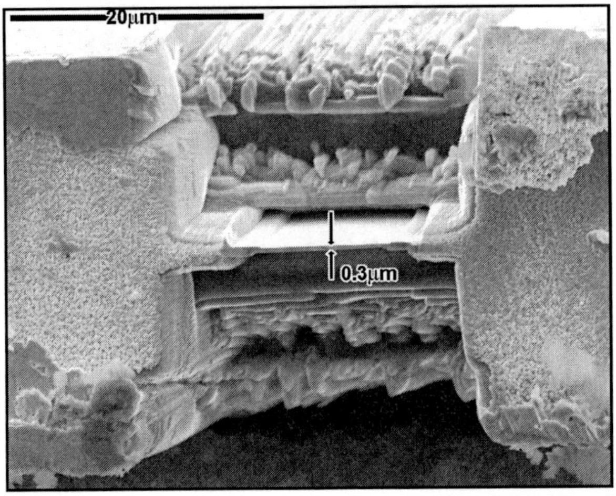

Figure 42: SEM image of a thin TEM sample milled by FIB. The thin membrane shown here is suitable for TEM examination; however, at ~300-nm thick, it would not be suitable for high-resolution TEM without further milling.

Ion etching is a sputtering process that can remove very fine quantities of material. This is used to perform a finishing polish of specimens polished by other means. Ion etching uses an inert gas passed through an electric field to generate a plasma stream that is directed to the sample surface. Acceleration energies for gases such as argon are typically a few kilovolts. The sample may be rotated to promote even polishing of the sample surface. The sputtering rate of such methods is on the order of tens of micrometers per hour, limiting the method to only extremely fine polishing.

More recently focused ion beam methods have been used to prepare samples. FIB is a relatively new technique to prepare thin samples for TEM examination from larger specimens. Because FIB can be used to micro-machine samples very precisely, it is possible to mill very thin membranes from a specific area of interest in a sample, such as a semiconductor or metal. Unlike inert gas ion sputtering, FIB makes use of significantly more energetic gallium ions and may alter the composition or structure of the material through gallium implantation.

Replication

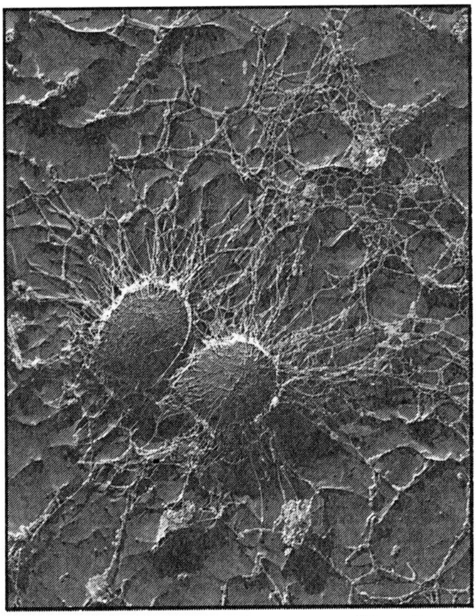

Figure 43: *Staphylococcus aureus* platinum replica image shot on a TEM at 50,000X magnification

Samples may also be replicated using cellulose acetate film, the film subsequently coated with a heavy metal, the original film melted away, and the replica imaged on the TEM. This technique is used for both materials and biological samples.

Modifications

The capabilities of the TEM can be further extended by additional stages and detectors, sometimes incorporated on the same microscope. An electron cryomicroscope (CryoTEM) is a TEM with a specimen holder capable of maintaining the specimen at liquid nitrogen or liquid helium temperatures. This allows imaging specimens prepared in vitreous ice, the preferred preparation technique for imaging individual molecules or macromolecular assemblies.

A TEM can be modified into a scanning transmission electron microscope (STEM) by the addition of a system that rasters the beam across the sample to form the image, combined with suitable detectors. Scanning coils are used to deflect the beam, such as by an electrostatic shift of the beam, where the beam is then collected using a current detector such as a faraday cup, which acts as a direct electron counter. By correlating the electron count to the position of the scanning beam (known as the "probe"), the transmitted component of the beam may be measured. The non-transmitted components may be obtained either by beam tilting or by the use of annular dark field detectors. In-situ experiments may also be conducted with experiments such as in-situ reactions or material deformation testing. Modern research TEMs may include aberration correctors, to reduce the amount of distortion in the image. Incident beam Monochromators may also be used which reduce the energy spread of the incident electron beam to less than 0.15 eV. Major TEM makers include JEOL, Hitachi High-technologies, FEI Company (from merging with Philips Electron Optics), Carl Zeiss and NION.

Low-voltage Electron Microscope (LVEM)

The low-voltage electron microscope (LVEM) is a combination of SEM, TEM and STEM in one instrument, which operated at relatively low electron accelerating voltage of 5 kV. Low voltage increases image contrast which is especially important for biological specimens. This increase in contrast significantly reduces, or even eliminates the need to stain. Sectioned samples generally need to be thinner than they would be for conventional TEM (20-65 nm). Resolutions of a few nm are possible in TEM, SEM and STEM modes.

Cryo-Electron Microscopy

This technique allows TEM's to be used to see molecular structure of proteins and large molecules. Cryoelectron microscopy involves viewing unaltered macromolecular assemblies by vitrifying them, placing them on a grid and obtaining images by detecting electrons that transmit through the specimen.

Limitations

There are a number of drawbacks to the TEM technique. Many materials require extensive sample preparation to produce a sample thin enough to be electron transparent, which makes TEM analysis a relatively time consuming process with a low throughput of samples. The structure of the sample may also be changed during the preparation process. Also the field of view is relatively small, raising the possibility that the region analysed may not be characteristic of the whole sample. There is potential that the sample may be damaged by the electron beam, particularly in the case of biological materials.

Resolution Limits

The limit of resolution obtainable in a TEM may be described in several ways, and is typically referred to as the information limit of the microscope. One commonly used value is a cut-off value of thecontrast transfer function, a function that is usually quoted in the frequency domain to define the reproduction of spatial frequencies of objects in the object plane by the microscope optics.

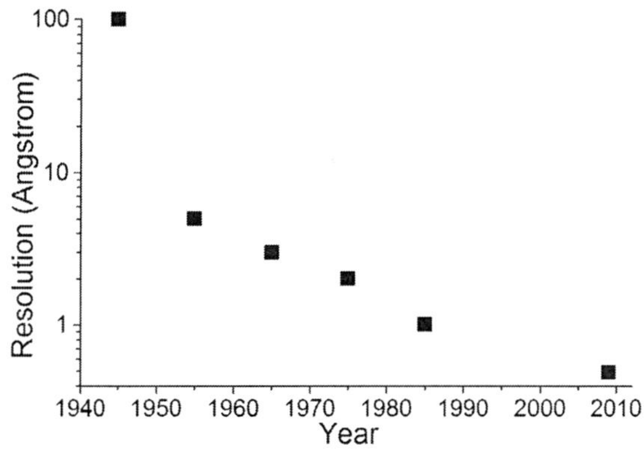

Figure 44: Approximate tendency of spatial resolution achieved with TEM.

A cut-off frequency, q_{max}, for the transfer function may be approximated with the following equation, where C_s is the spherical aberration coefficient and λ is the electron wavelength:

$$q_{\max} = \frac{1}{0.67(C_s\lambda^3)^{1/4}}.$$

For a 200 kV microscope, with partly corrected spherical aberrations ("to the third order") and a C_s value of 1 μm, a theoretical cut-off value might be $1/q_{max}$ = 42 pm. The same microscope without a corrector would have C_s = 0.5 mm and thus a 200-pm cut-off. The spherical aberrations are suppressed to the third or fifth order in the "aberration-corrected" microscopes. Their resolution is however limited by electron source geometry and brightness and chromatic aberrations in the objective lens system.

The frequency domain representation of the contrast transfer function may often have an oscillatory nature, which can be tuned by adjusting the focal value of the objective lens. This oscillatory nature implies that some spatial frequencies are faithfully imaged by the microscope, whilst others are suppressed. By combining multiple images with different spatial frequencies, the use of techniques such as focal series reconstruction can be used to improve the resolution of the TEM in a limited manner. The contrast transfer function can, to some extent, be experimentally approximated through techniques such as Fourier transforming images of amorphous material, such as amorphous carbon.

More recently, advances in aberration corrector design have been able to reduce spherical aberrations and to achieve resolution below 0.5 Engströms (50 pm) at magnifications above 50 million times. Improved resolution allows for the imaging of lighter atoms that scatter electrons less efficiently, such as lithium atoms in lithium battery materials.

The ability to determine the position of atoms within materials has made the HRTEM an indispensable tool for nanotechnology research and development in many fields, including heterogeneous catalysis and the development of semiconductor devices for electronics and photonics.

Chapter 7

Chemistry of Bioorganic Products

Organic Food

Organic foods are made in a way that complies with organic standards set by national governments and international organizations. In the United States, organic production is a system that is managed in accordance with the Organic Foods Production Act (OFPA) of 1990 and regulations in Title 7, Part 205 of the Code of Federal Regulations to respond to site-specific conditions by integrating cultural, biological, and mechanical practices that foster cycling of resources, promote ecological balance, and conserve biodiversity.

For the vast majority of human history, agriculture can be described as organic; only during the 20th century was a large supply of new synthetic chemicals introduced to the food supply. This more recent style of production is referred to as "conventional." Under organic production, the use of conventional non-organic pesticide (including insecticides, fungicides, and herbicides) is precluded.

However, contrary to popular belief, certain sprays and other materials that meet organic standards are allowed in the production of organic food. If livestock are involved, the livestock must be reared with regular access to pasture and without the routine use of antibiotics or growth hormones. In most countries, organic produce may not be genetically modified. It has been suggested that the application of nanotechnology to food and agriculture is a further technology that needs to be excluded from certified organic food. The Soil Association (UK) has been the first organic certifier to implement a nano-exclusion.

Organic food production is a heavily regulated industry, distinct from private gardening. Currently, the European Union, the United States, Canada, Japan and many other countries require producers to obtain special certification in order to market food as "organic" within their borders. Most certifications allow some chemicals and pesticides to be used, so consumers should be aware of the standards for qualifying as "organic" in their respective locales.

Historically, organic farms have been relatively small family-run operations, which is why organic food was once only available in small stores or farmers' markets. However, since the early 1990s organic food production has had growth rates of around 20% a year, far ahead of the rest of the food industry, in both developed and developing nations. As of April 2008, organic food accounts for 1–2% of food sales worldwide.

Meaning and Origin of the Term

In 1939, Lord Northbourne coined the term *organic farming* in his book *Look to the Land* (1940), out of his conception of "the farm as organism," to describe a holistic, ecologically-balanced approach to farming—in contrast to what he called *chemical farming*, which relied on "imported fertility" and "cannot be self-sufficient nor an organic whole." This is different from the scientific use of the term "organic," to refer to a class of molecules that contain carbon, especially those involved in the chemistry of life.

Environmental Impact

Several surveys and studies have attempted to examine and compare conventional and organic systems of farming. The general consensus across these surveys is that organic farming is less damaging for the following reasons:

- Organic farms do not consume or release synthetic pesticides into the environment—some of which have the potential to harm soil, water and local terrestrial and aquatic wildlife.
- Organic farms are better than conventional farms at sustaining diverse ecosystems, *i.e.*, populations of plants and insects, as well as animals.
- When calculated either per unit area or per unit of yield, organic farms use less energy and produce less waste, *e.g.*, waste such as packaging materials for chemicals.

However, some critics of organic farming methods believe that organic farms require more land to produce the same amount of food as conventional farms. They argue that if this is true, organic farms could potentially destroy the rain forests and wipe out many ecosystems.

A 2003 investigation by the Department for Environment Food and Rural Affairs in the UK found, similar to other reports, that organic farming "can produce positive environmental benefits", but that some of the benefits were decreased or lost when comparisons are made on "the basis of unit production rather than area".

Yield

One study found a 20% smaller yield from organic farms using 50% less fertilizer and 97% less pesticide. Studies comparing yields have had mixed results. Supporters claim that organically managed soil has a higher quality and higher water retention. This may help increase yields for organic farms in drought years.

One study from the Danish Environmental Protection Agency found that, area-for-area, organic farms of potatoes, sugar beet and seed grass produce as little as half the output of conventional farming. Findings like these, and the dependence of organic food on manure from low-yield cattle, has prompted criticism from scientists that organic farming is environmentally unsound and incapable of feeding the world population. Among these critics are Norman Borlaug, father of the "green revolution," and winner of the Nobel Peace Prize, who asserts that organic farming practices can at most feed 4 billion people, after expanding cropland dramatically and destroying ecosystems in the process. Michael Pollan, author of *The Omnivore's Dilemma*, responds to this by pointing out that the average yield of world agriculture is substantially lower than modern sustainable farming yields. Bringing average world yields up to modern organic levels could increase the worlds food supply by 50 %.

A 2007 study compiling research from 293 different comparisons into a single study to assess the overall efficiency of the two agricultural systems has concluded that...organic methods could produce enough food on a global per capita basis to sustain the current human population, and potentially an even larger population, without increasing the agricultural land base. The researchers also found that while in developed countries, organic systems on average produce 92% of the yield produced by conventional agriculture, organic systems produce 80% more than conventional farms in developing countries, because the materials needed for organic farming are more accessible than synthetic farming materials to farmers in some poor countries. On the other hand, communities that lack sufficient manure to replenish soils would struggle with organic farming, and the soil would degrade rapidly.

Energy Efficiency

A study of the sustainability of apple production systems showed that in comparing a conventional farming system to an organic method of farming, the organic system is more energy efficient. However, this is debatable due to organic farming's large use of tillage for weed control. Also increased fuel use from incorporating less nutrient dense fertilizers results in higher fuel consumption rates. The general analysis is that organic production methods are usually more energy efficient because they do not use chemically synthesized nitrogen. But they generally consume more petroleum because of the lack of other options for weed control and more intensive soil management practices. Energy efficiency is hard to determine; in the case listed above the author cites a book written in 1976. The true value of efficiency and energy consumption in relation to organic farms has yet to be determined.

Pesticides and Farmers

There are studies detailing the effects and side effects of pesticides upon the health of farm workers. Even when pesticides are organic, and are used correctly, they still end up in the air and bodies of farm workers. Through these studies, organophosphate pesticides have become associated with acute health problems such as abdominal pain, dizziness, headaches, nausea, vomiting, as well as skin and eye problems. In addition, there have been many other studies that have found pesticide exposure is associated with more severe health problems such as respiratory problems, memory disorders, dermatologic conditions, cancer, depression, neurologic deficits, miscarriages, and birth defects. Summaries of peer-reviewed research have examined the link between pesticide exposure and neurological outcomes and cancer in organophosphate-exposed workers. Those pesticides found to cause major health problems are banned for use in agriculture, conventional or organic, in many developed counties. Imported fruits and vegetables from Latin America are more likely to contain high level of pesticides, even pesticides banned for use in the United States. Migratory birds, such as Swainson's Hawks, have wintering grounds in Argentina where thousands of them were found dead from monocrotophos insecticide poisoning.

Safety and Pesticides

A study published by the National Research Council in 1993 determined that for infants and children, the major source of exposure to pesticides is through diet. A recent study in 2006 measured the levels of organophosphorus pesticide exposure in 23 schoolchildren before and after replacing their diet with organic food. In this study it was found that levels of organophosphorus pesticide exposure dropped dramatically and immediately when the children switched to an organic diet. Food residue limits established by law are set specifically with children in mind and consider a child's lifetime ingestion of each pesticide. There are controversial data on the health implications of certain pesticides. For example, the herbicide Atrazine has been, causing demasculinization in male frogs exposed to small concentrations. Under the effects of Atrazine, male frogs were found to have greatly increased occurrences of either malformed gonads, or testicular gonads which contain non-degenerate eggs. Effects were however significantly reduced in high concentrations, as is consistent with other teratogens affecting the endocrine system, such as estradiol.

Organic farming standards do not allow the use of synthetic pesticides, but they do allow the use of specific pesticides derived from plants. The most common organic pesticides, accepted for *restricted* use by most organic standards, include Bt, pyrethrum and rotenone. Rotenone has high toxicity to fish and aquatic creatures, causes Parkinson's disease if injected into rats, and shows other toxicity to mammals.

The United States Environmental Protection Agency and state agencies periodically review the licensing of suspect pesticides, but the process of de-listing is slow. One example of this slow process is exemplified by the pesticide Dichlorvos, or DDVP, which as recently as the year 2006 the EPA proposed its continued sale. The EPA has almost banned this pesticide on several occasions since the 1970s, but it never did so despite considerable evidence that suggests DDVP is not only carcinogenic but dangerous to the human nervous system—especially in children. The EPA "has determined that risks do not exceed levels of concern", a study of longterm exposure to DDVP in rats showed no toxic effects.

A study published in 2002 showed that "Organically grown foods consistently had about one-third as many residues as conventionally grown foods." Monitoring of pesticide residues in the United States is carried out by the Pesticide Data Program, a branch of the USDA created in 1990. It has since tested over 60 different types of food for over 400 different types of pesticides – with samples collected close to the point of consumption. Their most recent results found in 2005 that:

"These data indicate that 29.5 percent of all samples tested contained no detectable pesticides [parent compound and metabolite(s) combined], 30 percent contained 1 pesticide, and slightly over 40 percent contained more than 1 pesticide."

—USDA, Pesticide Data Program

Several studies corroborate this finding by having found that 25 percent of organic food carries synthetic pesticide residues, in comparison to 77 percent of conventional food.

Nutritional Value and Taste

In April 2009, results from Quality Low Input Food (QLIF), a 5-year integrated study funded by the European Commission, confirmed that "the quality of crops and livestock products from organic and conventional farming systems differs considerably." Specifically, results from a QLIF project studying the effects of organic and low-input farming on crop and livestock nutritional quality "showed that organic food production methods resulted in some case: (a) higher levels of nutritionally desirable compounds (e.g., vitamins/antioxidants and poly-unsaturated fatty acids such as omega-3 and CLA); (b) lower levels of nutritionally undesirable compounds such as heavy metals, mycotoxins, pesticide residues and glyco-alkaloids in a range of crops and/or milk; (c) a lower risk of faecal Salmonella shedding in pigs." but also showed no significant difference between traditionally grown foods on other studies.

The QLIF study also concludes that "further and more detailed studies are required to provide proof for positive health impacts of organic diets on human and animal health." Alternatively, according to the UK's Food Standards Agency, "Consumers may choose to buy organic fruit, vegetables and meat because they believe them to be more nutritious than other food.

However, the balance of current scientific evidence does not support this view." A 12-month systematic review commissioned by the FSA in 2009 and conducted at the London School of Hygiene & Tropical Medicine based on 50 years' worth of collected evidence concluded that "there is no good evidence that consumption of organic food is beneficial to health in relation to nutrient content." Other studies have found no proof that organic food offers greater nutritional values, more consumer safety or any distinguishable difference in taste. A recent review of nutrition claims showed that organic food proponents are unreliable information sources which harm consumers, and that consumers are wasting their money if they buy organic food believing that that it contains better nutrients. Regarding taste, a 2001 study concluded that organic apples were sweeter by blind taste test. Firmness of the apples was also rated higher than those grown conventionally. Current studies have not found differences in the amounts of natural biotoxins between organic and conventional foods.

Cost

Organic products typically cost 10 to 40% more than similar conventionally produced products. According to the USDA, Americans, on average, spent $1,347 on groceries in 2004; thus switching entirely to organics would raise their cost of groceries by $538.80 per year ($44.90/month) and switching to half organics would raise costs by $269.40 ($22.45/month). Processed organic foods vary in price when compared to their conventional counterparts. An Australian study by Choice magazine in 2004 found processed organic foods in supermarkets to be 65% more expensive, but noted this was not consistent. Prices may be higher because organic produce is produced on a smaller scale, and may need to be milled or processed separately. Furthermore, there is an increase in shipping costs from more centralized production in otherwise regional markets. In the case of dairy and eggs, the animal's requirements such as the number of animals that can be raised per acre, or the breed of animal and its feed conversion ratio affects the costs.

Related Movements

Biodynamic agriculture, a method of organic farming, is closely related to the organic food movement.

Food Co-ops and buyers groups are alternative ways of selling food which have also been closely related to the organic food movement. These co-operative structures have been aimed at lowering the retail price of organic food and some attempt to reduce the food miles associated with the produce.

Facts and Statistics

While organic food accounts for 1–2% of total food sales worldwide, the organic food market is growing rapidly, far ahead of the rest of the food industry, in both developed and developing nations.

- World organic food sales jumped from US $23 billion in 2002 to $52 billion in 2008.

- The world organic market has been growing by 20% a year since the early 1990s, with future growth estimates ranging from 10%–50% annually depending on the country.

Organic Clothing

Organic clothing is clothing made from materials raised or grown in compliance with organic agricultural standards.

Benefits

Authentic organic fabrics and clothing can help the environment in a number of ways, such as:

- Manufacture of chemicals is not required
- Chemical residues are not entered accidentally into the environment
- Humans and animals are not exposed to chemicals
- When the fabric is finished with chemicals are not returned to the earth in landfill, or enter into recycling process.

Cotton covers 2.5% of the world's cultivated land yet uses 16% of the world's insecticides, more than any other single major crop. It can take almost a 1/3 pound of synthetic fertilizers to grow one pound of raw cotton in the US, and it takes just under one pound of raw cotton to make one t-shirt.

Organic Cotton

Organic cotton is generally understood as cotton, from non genetically modified plants, that is to be grown without the use of any synthetic agricultural chemicals such as fertilizers or pesticides. Its production also promotes and enhances biodiversity and biological cycles. In the United States cotton plantations must also meet the requirements enforced by the National Organic Program (NOP), from the USDA, in order to be considered organic. This institution determines the allowed practices for pest control, growing, fertilizing, and handling of organic crops. As of 2007, 265,517 bales of organic cotton were produced in 24 countries and worldwide production was growing at a rate of more than 50% per year.

Ecological Footprint

Cotton covers 2.5% of the world's cultivated land yet uses 16% of the world's insecticides, more than any other single major crop. Other environmental consequences of the elevated use of chemicals in the non organic cotton growing methods consist of:

- High levels of agrochemicals are used in the production of non-organic, conventional cotton. Cotton production uses more chemicals per unit area than any other crop and accounts in total for 16% of the world's pesticides.
- Chemicals used in the processing of cotton pollute the air and surface waters.
- Residual chemicals may irritate consumers' skin.
- Decreased biodiversity and shifting equilibrium of ecosystems due to the use of pesticides.

Advantages

Cotton growers who make the transition to biologically based growing practices expect not only to offer a healthier and cleaner product, but also to benefit the planet. Some of the contributions to the different ecosystems include:

- Protecting surface and groundwater quality (eliminating contaminants in surface runoff)
- Reduced risk insect and disease control by replacing insecticide with the manipulation of ecosystems
- Long-term prevention of pests through beneficial habitat planting.
- Conservation of biodiversity
- Eliminate the use of toxic chemicals used in cotton
- Organically grown crops also yield soils with higher organic matter content, thicker topsoil depth, higher polysaccharide content, and lower modulus of rupture; therefore reducing considerably soil erosion.

Organic Certification

It is required by the law that any producer wanting to label and sell a product as "organic" must meet the standards established by the Organic Food Production Act of 1990, enforced by The State organic program (SOP) This act specifies the procedures and regulations for production and handling of organic crops.

Organic System Plan

Producers must elaborate an organic production or handling system plan which must also be approved by the state certifying agency or the USDA. This plan must include careful explanation of every process held in the plantation, as well as the frequency with which they are performed. A list of substances used on the crops is also necessary, along with a description of their composition, place where they will be used, and if possible documentation of commercial availability.

This inventory of substances is important for the regulation of allowed and prohibited material established by the SOP.Organic cotton growers must also provide A description of the control procedures and physical barriers established to prevent contact of organic and non organic crops on split operations and to avoid contact of organic production with prohibited substance during gestation, harvesting, and handling operations. This production plan can also be transferred to other states as long as it has already been approved by a certifying agency.

Production

Production requirements are specifically the set of changes that must be made to field and farming practices in order for a crop to be considered organic. To begin with, organic fields must go through a cleansing period of three years, without the use of any prohibited substances, before planting the first organic crop. Fields must also be equipped with physical barriers and buzzers in order to prevent contact of organic crops with any chemical substance product of surface runoff from crops near by. Producers must also strive to promote soil fertility through cultivation practices while maintaining or improving the physical, chemical, and biological condition of the soil and minimizes soil erosion. Organic growers must also implement practices to support biodiversity. Such practices include integrated pest management (IPM), which consists of the manipulation of ecosystems that benefit both the crops and the organisms that live around it. In addition to these practices, producers may only apply crop nutrients and soil amendments included on the National List of synthetic substances allowed in crop production.

Handling

Handling procedures are all the processes related to product packaging, pest control in handling processing facilities among others. The SOP allows the use of mechanical or biological methods for the purpose of retarding spoilage of products,but at the same time it prohibits the use of volatile synthetic solvents in processed products or any ingredient that is labelled as organic.

Pesticides

Since organic cotton is grown without the use of synthetic pesticides, it should contain fewer pesticides than conventional cotton. Pesticides used in the production of conventional cotton include orthophosphates such as phorate and methamidophos, endosulfan (highly toxic to farmers, but not very environmentally persistent) and aldicarb.. Other pesticides persisting in cotton fields in the United States include Trifluralin, Toxaphene and DDT. Although the last two chemicals are no longer used in the United States their long breakdown period and difficulty in removal ensures their persistence. Thus even organic cotton fields may contain them since conventional cotton fields can be transitioned to organic fields in 2–3 years.

Expanding Industry

Diverse institutions and campaigns are now educating the community about organic cotton and supporting growers on the switch to organic farming. The Sustainable Cotton Project is helping farmers in the transition from chemically dependent crops to more biological sound approaches. This institution has launched the Cleaner Cotton project, which promises to produce cotton with 73% less use of chemicals. The Community Alliance with Family Farmers (CAFF)also joined this effort since 2003. Their purpose is to educate farmers and the public about biological farming techniques and the importance of reducing chemical use in fiber and food production and supporting local farmers. In 2003, SCP joined the Community Alliance with Family Farmers (CAFF) to strengthen its operations and reach other farm and consumer audiences. CAFF and SCP provide growers with information about biological farming techniques and educate the public about the importance of reducing chemical use in fiber and food production and supporting local farmers.

Regional

Organic cotton is currently being grown successfully in many countries; the largest producers (as of 2007) are Turkey, India and China. Organic cotton production in Africa takes place in at least 8 countries. The earliest producer (1990) was the SEKEM organization in Egypt; the farmers involved later convinced the Egyptian government to convert 400,000 hectares of conventional cotton production to integrated methods, achieving a 90% reduction in the use of synthetic pesticides in Egypt and a 30% increase in yields.

Organic Baby Products

Organic baby products are made without pesticides, fertilizers, antibiotics, growth hormones, genetically modified organisms, additives, irradiation or sewage sludge. The manufacturer of such a product can only make these claims in marketing and advertising materials if the product has obtained organic certification. While organic baby products are becoming a big part of the organic movement, the upward trend has been met with some skepticism. However, organic advocacy groups such as the Organic Consumers Association work to educate the marketplace about the labelling system behind certified organic products.

Benefits

Organic baby products are substituted for conventional baby care items in order to reduce the amount of chemical exposure for rapidly developing infants. Some physicians recommend choosing organic foods for babies from conception until the age of three. During this period of time babies' bodies and brains grow faster and they are more vulnerable to environmental toxins. The fact that babies are smaller means they absorb more pesticides per pound of body weight than adults. The bottom line is toxins stay in their systems longer and can have more impact than fully developed adults. In addition to being free of pesticides and other additives, organic foods are believed to be more nutritional than their conventional counterparts. A 2007 European Union study confirmed that organic fruit and vegetables may have up to 40 percent more antioxidants than non-organically grown produce.

Disadvantages

Organic baby products are more expensive to produce and get certified. Therefore they are in most cases priced higher than conventional goods and are not as accessible to middle and lower income families. As a result, families that may believe there are health benefits to using organic baby products will not be able to afford them. The International Federation of Organic Agriculture Movements (IFOAM) says that while prices have been declining, there are reasons the costs of certified organic products remain higher than conventional products. They include limited supply as compared to demand, greater labour and transportation costs per unit produced. The smaller number of units involved also means marketing and distribution costs have to be spread out over a smaller number of units. As demand for organic food and products increases, technological innovations and economies of scale should reduce costs of production, processing, distribution and marketing for organic produce.

Products

Many factors are causing the accelerated growth of the organic baby product market. These include: increased frequency of baby product recalls for lead and other contaminants, more scientific studies finding toxic ingredients in baby care products, increased media exposure for the green and organic movement, the acknowledgment of the benefits of organics by some medical practitioners, and increased information sharing of the benefits in online mediums.

Baby Formula and Food

Mothers who are not breastfeeding can choose to use organic infant formula to meet their baby's nutritional requirements for the first four to six months of life. Certified organic baby formulas can be made from soy and dairy milk. Organic infant formulas can be obtained at the hospital, grocery store and through various channels on the internet. Parents can find organic versions of baby cereals in whole grain rice and whole grain oatmeal and multi-grain varieties. Some manufacturers fortify them with iron. Most brands make organic baby food in jars. The organic foods are blended to accommodate the various textures and blends of food desired by growing babies and toddlers.

Personal Care Products

A study reported in the February 2008 issue of *Pediatrics* found babies can be exposed to toxic chemicals in ordinary personal care products such as lotions, shampoos diaper creams. While the list of chemical ingredients involved in skin care manufacturing is extensive, this particular study focused in on phthalates, a reproductive toxin that can cause damage to sexual organs and impact future fertility. Another study released in March 2008 by the Organic Consumers Association found a cancer-causing chemical 1,4 Dioxane to be present in many personal care products.

Clothing

Baby clothing companies now offer items made from organically grown fibers, such as wool, cotton, hemp, soy or bamboo, that have been grown without the use of pesticides or produced without toxic fabric finishes and flame retardants. Herbal dyed organic clothing is the perfect clothing solution for infants, juveniles,babies and kids. These herbal dyed organic cotton clothes have medicinal value attached to it which nourishes kids' skin which is very soft and prone to allergies. These herbal clothes keep your baby's skin very soft and natural. These garments are dyed using medicinal rich herbs, plants, fruits, flowers and other natural minerals which are known to have medicinal/therapeutic properties like anti-allergic, anti-inflammatory.

Nursery Items

Mattresses and mattress pad covers should be made from organic cotton, organic wool and rubber latex from rubber-tree farms. The mattresses should be free of pesticides and other processing chemicals. Cloth diapers made from organic cotton, hemp or wool should be free of the chemical-based super absorbent gels, pesticides and chlorine found in conventional diapers. Sleepwear/Pajamas made from organic cotton should be free of pesticides, flame retardants, and synthetics, such as polyester and acrylic. They must be snug-fitting if they have not been treated with flame retardant chemicals. Certified organic toys have no pesticides and contain no toxic dyes and paints. Toys should be made from organic fabrics and non-toxic stuffings; wooden toys should use water- or milk-based paints and stained with natural oils or beeswax. Upholstered nursery furniture should be made from untreated, sustainable wood and covered in organic fabrics to reduce the amount of toxins in the nursery.

Organic Fertilizer

Organic fertilizers are naturally-occurring fertilizers (e.g. peat moss or green manure), or naturally occurring mineral deposits (e.g. saltpeter). Naturally occurring organic fertilizers include manure, slurry, worm castings, peat, seaweed, humic acid, brassin and guano. Sewage sludge use in organic agricultural operations in the U.S. has been extremely limited and rare due to USDA prohibition of the practice (due to toxic metal accumulation, among other factors). Processed organic fertilizers include compost, bloodmeal, bone meal, humic acid, amino acids, brassin and seaweed extracts. Other examples are natural enzyme digested proteins, fish meal, and feather meal. Decomposing crop residue from prior years is another source of fertility.

Discussion of the Term 'Organic'

The notion of organic growing started out as a scientific process.

Organic systems are based on the element, Carbon. Carbon is essential to any compost process.

In later years, "organic" became a legal term. For example, there used to be a distinction between the term "organic" and the term "pesticide free". Organic simply dealt with the use of fertilizer types. Once the term "organic" became regulated, many other factors were added. "Pesticide free" is not at all related to fertilization (plant nutrition), but has become a legal inclusion. Likewise, in scientific terms, a fish emulsion can be a good organic fertilizer, but in some jurisdictions fish emulsion must be certified "dolphin safe" to be considered "organic".

Natural Sourcing

Animal-sourced Urea and Urea-Formaldehyde (from urine), are suitable for application organic agriculture, while pure synthetic forms are not deemed, however, pure (synthetically-produced) urea is not. The common thread that can be seen through these examples is that *organic* agriculture attempts to define itself through minimal processing (e.g. via chemical energy such as petroleum), as well as being naturally-occurring or via natural biological processes such as composting. Cover crops are also grown to enrich soil as a green manure through nitrogen fixation from the atmosphere; as well as phosphorus (through nutrient mobilization) content of soils. Powdered limestone, mined rock phosphate and Chilean saltpeter, are *inorganic* chemicals in the technical (organic chemistry) sense of the word, but are considered suitable for organic agriculture..

Advantages

Although the density of nutrients in organic material is comparatively modest, they have many advantages. The majority of nitrogen supplying organic fertilizers contain insoluble nitrogen and act as a slow-release fertilizer. By their nature, organic fertilizers increase physical and biological nutrient storage mechanisms in soils, mitigating risks of over-fertilization. Organic fertilizer nutrient content, solubility, and nutrient release rates are typically much lower than mineral (inorganic) fertilizers. A University of North Carolina study found that potential mineralizable nitrogen (PMN) in the soil was 182–285% higher in organic mulched systems, than in the synthetics control.

Organic fertilizers also re-emphasize the role of humus and other organic components of soil, which are believed to play several important roles:

- Mobilizing existing soil nutrients, so that good growth is achieved with lower nutrient densities while wasting less

- Releasing nutrients at a slower, more consistent rate, helping to avoid a boom-and-bust pattern
- Helping to retain soil moisture, reducing the stress due to temporary moisture stress
- Improving the soil structure
- Helping to prevent topsoil erosion (responsible for desertfication and the Dust bowl.

Organic fertilizers also have the advantage of avoiding certain problems associated with the regular heavy use of artificial fertilizers:

- The necessity of reapplying artificial fertilizers regularly (and perhaps in increasing quantities) to maintain fertility
- Extensive runoff of soluble nitrogen and phosphorus, leading to eutrophication of bodies of water (which causes fish kills)
- Costs are lower for if fertilizer is locally available.

According to the PPI institute website, it is widely thought that organic fertilizer is better than inorganic fertilizer. However, balanced responsible use of either or both can be just as good for the soil.

Disadvantages

Organic fertilizers have the following disadvantages:

- As a dilute source of nutrients when compared to inorganic fertilizers, transporting large amount of fertilizer incurs higher costs, especially with slurry and manure.
- The composition of organic fertilizers tends to be more complex and variable than a standardized inorganic product.
- *Improperly*-processed organic fertilizers may contain pathogens from plant or animal matter that are harmful to humans or plants. However, proper composting should remove them.
- More labour is needed to compost organic fertilizer, increasing labour costs. Some of this cost is offset by reduced cash purchase.

Conventional Farming Application

In non-organic farming a compromise between the use of artificial and organic fertilizers is common, often using inorganic fertilizers supplemented with the application of organics that are readily available such as the return of crop residues or the application of manure.

Defining Large Proteins and M13 Coat Proteins

M13 is a filamentous bacteriophage composed of circular single stranded DNA (ssDNA) which is 6407 nucleotides long encapsulated in approximately 2700 copies of the major coat protein P8, and capped with 5 copies of two different minor coat proteins (P9, P6, P3) on the ends. The minor coat protein P3 attaches to the receptor at the tip of the F pilus of the host *Escherichia coli*. Infection with filamentous phages is not lethal, however the infection causes turbid plaques in *E. coli*. It is a non-lytic virus. However a decrease in the rate of cell growth is seen in the infected cells. M13 plasmids are used for many recombinant DNA processes, and the virus has also been studied for its uses in nanostructures and nanotechnology.

Phage Particles

The phage coat is primarily assembled from a 50 amino acid protein called pVIII (or p8), which is encoded by gene VIII (or g8) in the phage genome. For a wild type M13 particle, it takes about approximately 2700 copies of p8 to make the coat about 900 nm long. The coat's dimensions are flexible though and the number of p8 copies adjusts to accommodate the size of the single stranded genome it packages. For example, when the phage genome was mutated to reduce its number of DNA bases (from 6.4 kb to 221 bp), then the number of p8 copies was decreased to fewer than 100, causing the p8 coat to shrink in order to fit the reduced genome. The phage appear to be limited at approximately twice the natural DNA content. However, deletion of a phage protein (p3) prevents full escape from the host *E. coli*, and phage that are 10-20X the normal length with several copies of the phage genome can be seen shedding from the *E. coli* host.

There are four other proteins on the phage surface, two of which have been extensively studied. At one end of the filament are five copies of the surface exposed pIX (p9) and a more buried companion protein, pVII (p7). If p8 forms the shaft of the phage, p9 and p7 form the "blunt" end that is seen in the micrographs. These proteins are very small, containing only 33 and 32 amino acids respectively, though some additional residues can be added to the N-terminal portion of each which are then presented on the outside of the coat. At the other end of the phage particle are five copies of the surface exposed pIII (p3) and its less exposed accessory protein, pVI (p6). These form the rounded tip of the phage and are the first proteins to interact with the *E. coli* host during infection. p3 is also the last point of contact with the host as new phage bud from the bacterial surface.

Phage Life-cycle

The general stages to a viral life cycle are: infection, replication of the viral genome, assembly of new viral particles and then release of the progeny particles from the host. Filamentous phage use a bacterial structure known as the F pilus to infect *E. coli*, with the M13 p3 tip contacting the TolA protein on the bacterial pilus. The phage genome is then transferred to the cytoplasm of the bacterial cell where resident proteins convert the single stranded DNA genome to a double stranded replicative form ("RF"). This DNA then serves as a template for expression of the phage genes.

Two phage gene products play critical roles in the next stage of the phage life cycle, namely amplification of the genome. pII (aka p2) nicks the double stranded form of the genome to initiate replication of the + strand. Without p2, no replication of the phage genome can occur. Host enzymes copy the replicated + strand, resulting in more copies of double stranded phage DNA. pV (aka p5) competes with double stranded DNA formation by sequestering copies of the + stranded DNA into a protein/DNA complex destined for packaging into new phage particles. Interestingly there is one additional phage-encoded protein, pX (p10), that is important for regulating the number of double stranded genomes in the bacterial host. Without p10 no + strands can accumulate. What's particularly interesting about p10 is that it's identical to the C-terminal portion of p2 since the gene for p10 is within the gene for p2 and the protein arises from transcription initiation within gene 2. This makes the manipulation of p10 inextricably linked to manipulation of p2 (an engineering headache) but it also makes for a compact and efficient phage in nature.

Phage maturation requires the phage-encoded proteins pIV (p4), pI (p1) and its translational restart product pXI (p11). Multiple copies (on the order of 12 or 14) of p4 assemble in the outer membrane into a stable, i.e. detergent resistant, barrel-shaped structure. Similarly a handful of the p1 and p11 proteins (5 or 6 copies of each) assemble in the bacterial inner membrane, and genetic evidence suggests C-terminal portions of p1 and p11 interact with the N-terminal portion of p4 in the periplasm. Together the p1, p11, p4 complex forms channels through which mature phage are secreted from the bacterial host.

To initiate phage secretion, two of the minor phage coat proteins, p9 and p7, are thought to interact with the p5-single stranded DNA complex at a region of the DNA called the packaging sequence (aka PS). The p5 proteins covering the single stranded DNA are then replaced by p8 proteins that are embedded in the bacterial membrane and the growing phage filament is threaded through the p1, p11, p4 channel. This replacement of p5 by p8 explains the microphage data presented earlier indicate how the size of the phage particle is determined by the number of bases the phage packages.

Once the phage DNA has been fully coated with p8, the secretion terminates by adding the p3/p6 cap, and the new phage detaches from the bacterial surface. How long does all this take? Amazingly, new M13 'phage particles are secreted within 10 minutes from a newly infected host and can arise at a rate of 1000/cell within the first hour of infection. The bacterial host can continue to grow and divide, allowing this process to continue indefinitely.

Phage proteins in the cytoplasm are pII, pX, and pV, and they are part of the replication process of DNA. The other phage proteins are synthesized and inserted into the cytoplasmic or outer membranes.

- pV dimers bind newly synthesized single-stranded DNA and prevent conversion to RF DNA
- RF DNA synthesis continues and amount of pV reaches critical concentration
- DNA replication switches to synthesis of single-stranded (+) viral DNA
- pV-DNA structures from about 800 nm long and 8 nm in diamter
- pV-DNA complex is substrate in phage assembly reaction.

Helices and Metallacycles

A helix (pl: helixes or helices) is a type of space curve, i.e. a smooth curve in three-dimensional space. It is characterized by the fact that the tangent line at any point makes a constant angle with a fixed line called the *axis*. Examples of helixes are coil springs and the handrails of spiral staircases. A "filled-in" helix – for example, a spiral ramp – is called a helicoid. Helices are important in biology, as the DNA molecule is formed as two intertwined helices, and many proteins have helical substructures, known as alpha helices. The word *helix* comes from the Greek.

Types

Helices can be either right-handed or left-handed. With the line of sight along the helix's axis, if a clockwise screwing motion moves the helix away from the observer, then it is called a right-handed helix; if towards the observer then it is a left-handed helix. Handedness (or chirality) is a property of the helix, not of the perspective: a right-handed helix cannot be turned or flipped to look like a left-handed one unless it is viewed in a mirror, and vice versa. Most hardware screws are right-handed helices. The alpha helix in biology as well as the A and B forms of DNA are also right-handed helices. The Z form of DNA is left-handed.

The pitch of a helix is the width of one complete helix turn, measured parallel to the axis of the helix. A double helix consists of two (typically congruent) helices with the same axis, differing by a translation along the axis, which may or may not measure half the pitch. A conic helix may be defined as a spiral on a conic surface, with the distance to the apex an exponential function of the angle indicating direction from the axis. An example is the Corkscrew roller coaster at Cedar Point amusement park. A circular helix has constant band curvature and constant torsion. A curve is called a general helix or cylindrical helix if its tangent makes a constant angle with a fixed line in space. A curve is a general helix if and only if the ratio of curvature to torsion is constant.

Mathematical Description

In mathematics, a helix is a curve in 3-dimensional space. The following parametrisation in Cartesian coordinates defines a helix:

$$x(t) = \cos(t),$$
$$y(t) = \sin(t),$$
$$z(t) = t.$$

As the parameter t increases, the point $(x(t),y(t),z(t))$ traces a right-handed helix of pitch 2π and radius 1 about the z-axis, in a right-handed coordinate system.

In cylindrical coordinates (r, θ, h), the same helix is parametrised by:

$$r(t) = 1,$$
$$\theta(t) = t,$$
$$h(t) = t.$$

A circular helix of radius a and pitch $2\pi b$ is described by the following parametrisation:

$$x(t) = a\cos(t),$$
$$y(t) = a\sin(t),$$
$$z(t) = bt.$$

Another way of mathematically constructing a helix is to plot a complex valued exponential function taking imaginary arguments.

Except for rotations, translations, and changes of scale, all right-handed helices are equivalent to the helix defined above. The equivalent left-handed helix can be constructed in a number of ways, the simplest being to negate any one of the x, y or z components.

Arc Length, Curvature and Torsion

The length of a circular helix of radius a and pitch 2πb expressed in rectangular coordinates as

$$t \mapsto (a\cos t, a\sin t, bt), t \in [0, T]$$

equals $T\cdot\sqrt{a^2+b^2}$ its curvature is $\frac{|a|}{a^2+b^2}$ and its torsion is $\frac{b}{a^2+b^2}$.

Examples

In music, pitch space is often modeled with helices or double helices, most often extending out of a circle such as the circle of fifths, so as to represent octave equivalency.

In organometallic chemistry, a metallacycle is a derivative of a carbocyclic compound wherein a metal has replaced at least one carbon centre. Metallacycles appear frequently as reactive intermediates in catalysis, e.g. olefin metathesis and alkyne trimerization. In organic synthesis, directed ortho metalation is widely used for the functionalization of arene rings via C-H activation. One main effect that metallic atom substitution on a cyclic carbon compound is distorting the geometry due to the large size of typical metals.

Nomenclature

A wide variety of cyclic compounds containing metals in rings are known, e.g. all chelate rings. Usually, such compounds are not classified as metallacycles, but the naming conventions are not rigidly followed. Within the area of coordination chemistry and supramolecular chemistry, examples include metallacrowns, metallacryptands, metallahelices, and molecular wheels.

Classes of Metallacycles

Metal-alkene complexes can be viewed as the smallest metallacycles, but they usually are not classified as metallacycles. In the Dewar-Chatt-Duncanson model, one resonance structure for the M(η^2-alkene) centre is the metallacyclopropane.

Metallacyclobutanes

The parent metallacyclobutane has the formula $L_nM(CH_2)_3$. A stable example is ([[triphenylphosphine | $PPh_3)_2Pt(CH_2)_3$.

Metallacyclobutane intermediates are involved in the alkene metathesis and in the oligomerization and dimerization of ethylene. Relevant to alkene metathesis, the Chauvin mechanism invokes the attack of an alkene at an electrophilic metal carbene catalyst. This work helped to validate the Chauvin mechanism for olefin metathesis.

Metallacyclopentadienes

The parent metallacyclopentadiene has the formula $L_nM(CH)_4$. Most arise from the coupling of two alkynes at a low valent metal centres such as derivatives of Co(I) and Zr(II). Late metal derivatives (Co, Ni) are intermediates in the metal-catalysed trimerization of alkynes to arenes. Early metal derivatives, i.e. derivatives of Ti and Zr, are used stoichiometrically. For example, the zirconacyclopentadiene $Cp_2ZrC_4Me_4$ is a useful carrier for $C_4Me_4^{2-}$.

Metallacyclopentanes

The parent metallacyclopentane has the formula $L_nM(CH_2)_4$. Such compounds are intermediates in the metal catalysed dimerization, trimerization, and tetramerization of ethylene to give 1-butene, 1-hexene, and 1-octene, respectively.

Metallabenzenes

The parent metallacyclobutane has the formula $L_nM(CH)_5$. Since the discovery of the structure of benzene, many analogues of this iconic structure have been described. Examples include pyridine, phosphabenzene, arsabenzene, and pyrylium. Included in this group are the metallabenzenes.

Metallabenzene complexes have been classified into three varieties; in such compounds the parent acyclic hydrocarbon ligand is viewed as the anion $C_5H_5^-$. The 6 π electrons in the metallacycle conform to the Hückel (4n+2) theory.

Isolated and characterized metallabenzenes have been also been reported with metals ruthenium, iridium, platinum and rhenium.

Ortho-metallation

Metallacycles often arise by cyclization of arene-containing donor ligands, e.g. aryl phosphines and amines. An early example is the cyclization of $IrCl(PPh_3)_3$ to give the corresponding Ir(III) hydride containing a four-membered IrPCC ring. Palladium(II) and platinum(II) have long been known to ortho-metallate aromatic ligands such as azobenzene, benzylamines, and 2-phenylpyridines. These reactions are strongly influenced by substituent effects, including the Thorpe-Ingold Effect. Ligands that lack aryl substituents will sometimes cyclometalate via activation of methyl groups, an early example being the internal oxidative addition of methylphosphine ligands. Metallacycle formation interferes with intermolecular C-H activation processes. For this reason, specialized "pincer ligands" ligands have been developed that resist ortho-metallation.

Copper is a very important trace element in the human body. It takes part in a lot of biological functions such as the hemoglobin synthesis, pigmentation, bone formation, reproduction, fertility development and proper functioning of the central and peripheral nervous systems. These functions of copper are carried out in association with copper-dependent enzymes. For example, dopamine-β-hydroxylase, which is necessary for the synthesis of norepinephrine, serves for the metabolic function; the cytochrome *C* oxidase helps the cellular utilization of oxygen; the superoxide dismutase is used in the disproportionation of superoxide and soluble pyridoxal dependent monoamine oxidase assists in the oxidation of catecholamine to aldehydes.

The suitable redox potential of Cu(II) in the solution makes it play an important role in the electron transition processes in the biological systems. Furthermore, there are a number of cases where copper is needed for, or enhances the biological activity of potential ligand systems. For example, the copper(II) complexes of the *N*methylisatin β-thiosemicarbazone are much more active than the free ligand. The biological activities of other synthesized Cu(II) complexes have also been known for a long time.

Besides their biological activities, copper complexes also exhibit a number of interesting structures such as helices and metallacycles by self-assembly and recognition process. Because of its orbital configuration, Cu(II) provides a typical example of Jahn-Teller effect. Copper centres in the complexes have a variety of distortions, which have attracted the attention of chemists for a few decades.

Geometry of Cu(II) Complexes

The Cu(II) ion is the most common oxidation state of copper exists in biological systems. It is considered hard, and easily to coordinates with N and O. The coordination number is mainly four, five or six. When Cu(II) is four coordinated, the geometry of the metal centre is square planar. When Cu(II) is five coordinated, the geometry of the metal centre is square distorted pyramidal or trigonal bipyramidal. When Cu(II) is six coordinated, the geometry of the metal centre is elongated octahedron. Cu(II) may also have intermediate structures between square pyramidal and trigonal bipyramidal geometry when it is five coordinated. The judgment on the five coordinated geometry between square pyramid and trigonal bipyramid can be made based on the τ value, which was introduced by Reedijk and co-workers. It is defined as $\tau = (\beta\text{-}\alpha)/60$, which is the ratio differences between two in-plane angles, that are larger than 90°. Normally, the τ value is between 0 and 1. The absolute number 0 represents the perfect square pyramid ($C4$v) and 1 represents the ideal trigonal bipyramidal ($D3$h). The ideal situation is seldom seen. This definition has been widely used to compare the structure of five coordinate Cu(II) complexes.

Copper (II) Complexes with Schiff Base Analogues

Considerable attention has been given to Schiff base adducts formed between pyridoxal or analogues and amino acids. Studies in pyridoxal analogues started from the interest in their biological activities. Pyridoxal is Vitamin B6, which is commonly regarded as one of the most important growth factors for both animals and humans. It was found that pyridoxal catalyses a variety of amino acid reactions. Pyridoxal 5'-phosphate, one of the most important pyridoxal analogues, acts as a principal coenzyme in almost all reactions of the primary amino acids. Many reactions are catalysed by pyridoxal 5'-phosphate containing enzymes. These include transamination, decarboxylation and racemisation. The initial combination of pyridoxal phosphate with amino acids to form a Schiff base is a common feature of these two components dependent amino acid transformations. Thus a number of pyridoxal Schiff base compounds and their analogues, with a range of amino acids have previously been synthesized.37,38 The metal complexes of Schiff base derivatives have been shown to be a useful model for more complicated metal – pyridoxal-amino acid system, which are intermediates in biologically important transmission reactions.

Texaphyrin-type Expanded Porphyrins

Texaphyrin is a sub-class of heterocyclic macrocycle molecules known as porphyrins. The molecule was invented by University of

Texas at Austin professor Jonathan Sessler. The name texaphryin arose because some of the molecules have a shape that can superimpose onto the points of the star featured on the state flag of Texas. Texaphyrins were nominated as the "State Molecule of Texas", but the buckyball was chosen instead.

Sessler has described possible medicinal uses of these compounds in the *Proceedings of the National Academy of Sciences* and other scientific journals. Pharmacyclics, Inc., a publicly-traded company begun by Sessler, licensed the technology behind texaphyrins from the university in order to develop commercial and medical uses for the molecules. One is being developed under the name Antrin. Two predicted applications will be in neoplastic and cardiovascular disease.

Porphyrin

Porphyrins are a group of organic compounds of which many occur in nature. One of the best-known porphyrins is heme, the pigment in red blood cells. Heme is a cofactor of the protein hemoglobin. They are heterocyclic macrocycles composed of four modified pyrrole subunits interconnected at their α carbon atoms via methine bridges (=CH-). Porphyrins are aromatic. That is, they obey Hückel's rule for aromaticity, possessing $4n+2 = \pi$ electrons (n=4 for the shortest cyclic path) that are delocalized over the macrocycle. The macrocycles, therefore, are highly-conjugated systems. As a consequence, they typically have very intense absorption bands in the visible region and may be deeply coloured; the name porphyrin comes from a Greek word for *purple*. The macrocycle has 26 pi electrons in total. The parent porphyrin is porphine, and substituted porphines are called porphyrins.

Complexes of Porphyrins and Related Molecules

Porphyrins are the conjugate acids of ligands that bind metals to form complexes. The metal ion usually has a charge of 2+ or 3+. A schematic equation for these syntheses is shown:

$$H_2\text{porphyrin} + [ML_n] \rightarrow M(\text{porphyrinate})L_{n-4} + 4\ L + 2\ H^+$$

A porphyrin without metal in its cavity is a *free base*. Some iron-containing porphyrins are called hemes. Heme-containing proteins, or *hemoproteins*, are found extensively in nature. Hemoglobin and myoglobin are two O_2-binding proteins that contain iron porphyrins.

Several other heterocycles are related to porphyrins. These include corrins, chlorins, bacteriochlorophylls, and corphins. Chlorins (2,3-dihydroporphyrin) are more reduced, contain more hydrogen than porphyrins, and feature a pyrroline subunit. This structure occurs in chlorophyll. Replacement of two of the four pyrrolic subunits with pyrrolinic subunits results in either a bacteriochlorin (as found in some photosynthetic bacteria) or an isobacteriochlorin, depending on the relative positions of the reduced rings. Some porphyrin derivatives follow Hückel's rule, but most do not.

Synthesis

Biosynthesis: The "committed step" for porphyrin biosynthesis is the formation of D-aminolevulinic acid (dALA) by the reaction of the amino acid glycine with succinyl-CoA from the citric acid cycle. Two molecules of dALA combine to give porphobilinogen (PBG), which contains a pyrrole ring. Four PBGs are then combined through deamination into hydroxymethyl bilane (HMB), which is hydrolysed to form the circular tetrapyrrole uroporphyrinogen III. This molecule undergoes a number of further modifications. Intermediates are used in different species to form particular substances, but, in humans, the main end-product protoporphyrin IX is combined with iron to form heme. Bile pigments are the breakdown products of heme.

Laboratory Synthesis

One of the more common syntheses for porphyrins is based on work by Paul Rothemund. His techniques underpin more modern syntheses such as those described by Adler and Longo. The synthesis of simple porphyrins such as *meso*-tetraphenylporphyrin (H_2TPP) is also commonly done in university teaching labs. It can be done by heating an equimolar mixture of pyrrole and benzaldehyde in refluxing propionic acid for about an hour. In this method, porphyrins are assembled from pyrrole and substituted aldehydes. Acidic conditions are essential; formic acid, acetic acid, and propionic acid are typical reaction solvents, or p-toluenesulfonic acid can be used with a non-acidic solvent. Lewis acids such as boron trifluoride etherate and ytterbium triflate have also been known to catalyse porphyrin formation. A large amount of side-product is formed and is removed, usually by recrystallization or chromatography.

Applications

Although natural porphyrin complexes are essential for life, synthetic porphyrins and their complexes have limited utility. Complexes of meso-tetraphenylporphyrin, e.g., the iron(III) chloride complex (TPPFeCl), catalyse a variety of reactions in organic synthesis, but none of them are of practical value. Porphyrin-based compounds are of interest in molecular electronics and supramolecular building blocks. Phthalocyanines, which are structurally related to porphyrins, are used in commerce as dyes and catalysts. Synthetic porphyrin dyes that are incorporated in the design of solar cells are the subject of ongoing research.

In 2008, the corporation Destiny Pharma reported successful clinical trials of an intra-nasally applied porphyrin XF-73 against methicillin-resistant *Staphylococcus aureus*.

Supramolecular Chemistry

Porphyrins are often used to construct structures in supramolecular chemistry. These systems take advantage of the Lewis acidity of the metal, typically zinc. An example of a host-guest complex that was constructed from a macrocycle composed of four porphyrins. A guest-free base porphyrin is bound to the centre by coordination with its four-pyridine substituents.

Organic Geochemistry

The field of organic geochemistry, the study of the impacts and processes that organisms have had on the Earth, had its origins in the isolation of porphyrins from petroleum. This finding helped establish the biological origins of petroleum. Petroleum is sometimes "fingerprinted" by analysis of trace amounts of nickel and vanadyl porphyrins.

Chlorophyl is a magnesium porphyrin, and heme is an iron porphyrin, but neither mineral is present in petroleum. On the other hand, nickel and vanadyl porphyrins could be related to catalytic molecules from bacteria that feed primordial hydrocarbons.

A Porphyrin-related Disease: Porphyria

Porphyrias are a group of inherited or acquired disorders of certain enzymes in the heme bio-synthetic pathway (also called porphyrin pathway). They are broadly classified as acute (hepatic) porphyrias and cutaneous (erythropoietic) porphyrias, based on the site of the overproduction and accumulation of the porphyrins (or their chemical precursors). They manifest with either neurological complications or skin problems (or occasionally both). A clinically induced and histologically identical condition is called pseudoporphyria. Pseudoporphyria is characterized by normal serum and urine porphyrin levels.

The name is likely to have been a reference to the purple discolouration of feces and urine in patients during an attack. Although original descriptions are attributed to Hippocrates, the disease was first explained biochemically by Dr Felix Hoppe-Seyler in 1874, and acute porphyrias were described by the Dutch physician Prof B.J. Stokvis in 1889.

Signs and Symptoms

Acute Porphyria

The acute, or hepatic, porphyrias primarily affect the nervous system, resulting in abdominal pain, vomiting, acute neuropathy, muscle weakness, seizures, and mental disturbances, including hallucinations, depression, anxiety, and paranoia. Cardiac arrhythmias and tachycardia (fast heart rate) may develop as the autonomic nervous system is affected. Pain can be severe and can, in some cases, be both acute and chronic in nature. Constipation is frequently present, as the nervous system of the gut is affected, but diarrhea can also occur.

Given the many presentations and the relatively uncommon occurrence of porphyria, the patient may initially be suspected to have other, unrelated conditions. For instance, the polyneuropathy of acute porphyria may be mistaken for Guillain-Barré syndrome, and porphyria testing is commonly recommended in those scenarios. Systemic lupus erythematosus features photosensitivity and pain attacks and shares various other symptoms with porphyria.

Not all porphyrias are genetic, and patients with liver disease who develop porphyria as a result of liver dysfunction may exhibit other signs of their condition, such as jaundice.

Patients with acute porphyria (AIP, HCP, VP) are at increased risk over their life for hepatocellular carcinoma (primary liver cancer) and may require monitoring. Other typical risk factors for liver cancer need not be present.

Cutaneous Porphyria

The cutaneous, or erythropoietic, porphyrias primarily affect the skin, causing photosensitivity (photodermatitis), blisters, necrosis of the skin and gums, itching, and swelling, and increased hair growth on areas such as the forehead. Often there is no abdominal pain, distinguishing it from other porphyrias.

In some forms of porphyria, accumulated heme precursors excreted in the urine may cause various changes in colour, after exposure to sunlight, to a dark reddish or dark brown colour. Even a purple hue or red urine may be seen.

Diagnosis

Porphyrin Studies

Porphyria is diagnosed through spectroscopy and biochemical analysis of blood, urine, and stool. In general, urine estimation of porphobilinogen (PBG) is the first step if acute porphyria is suspected. As a result of feedback, the decreased production of heme leads to increased production of precursors, PBG being one of the first substances in the porphyrin synthesis pathway. In nearly all cases of acute porphyria syndromes, urinary PBG is markedly elevated except for the very rare ALA dehydratase deficiency or in patients with symptoms due to hereditary tyrosinemia type I. In cases of mercury- or arsenic poisoning-induced porphyria, other changes in porphyrin profiles appear, most notably elevations of uroporphyrins I & III, coproporphyrins I & III and pre-coproporphyrin.

Repeat testing during an attack and subsequent attacks may be necessary in order to detect a porphyria, as levels may be normal or near-normal between attacks. The urine screening test has been known to fail in the initial stages of a severe life threatening attack of acute intermittent porphyria.

The bulk (up to 90%) of the genetic carriers of the more common, dominantly inherited acute hepatic porphyrias (acute intermittent porphyria, hereditary coproporphyria, variegate porphyria) have been noted in DNA tests to be latent for classic symptoms and may require DNA or enzyme testing. The exception to this may be latent post-puberty genetic carriers of hereditary coproporphyria.

As most porphyrias are rare conditions, general hospital labs typically do not have the expertise, technology or staff time to perform porphyria testing. In general, testing involves sending samples of blood, stool and urine to a reference laboratory. All samples to detect porphyrins must be handled properly. Samples should be taken during an acute attack, otherwise a false negative result may occur. Samples must be protected from light and either refrigerated or preserved. If all the porphyrin studies are negative, one has to consider pseudoporphyria. A careful medication review often will find the inciting cause of pseudoporphyria.

Additional Tests

Further diagnostic tests of affected organs may be required, such as nerve conduction studies for neuropathy or an ultrasound of the liver. Basic biochemical tests may assist in identifying liver disease, hepatocellular carcinoma, and other organ problems.

Pathogenesis

In humans, porphyrins are the main precursors of heme, an essential constituent of hemoglobin, myoglobin, catalase, peroxidase, respiratory and P450 liver cytochromes.

Deficiency in the enzymes of the porphyrin pathway leads to insufficient production of heme. Heme function plays a central role in cellular metabolism. This is not the main problem in the porphyrias; most heme synthesis enzymes—even dysfunctional enzymes—have enough residual activity to assist in heme biosynthesis. The principal problem in these deficiencies is the accumulation of porphyrins, the heme precursors, which are toxic to tissue in high concentrations. The chemical properties of these intermediates determine the location of accumulation, whether they induce photosensitivity, and whether the intermediate is excreted (in the urine or feces).

There are eight enzymes in the heme biosynthetic pathway, four of which—the first one and the last three—are in the mitochondria, while the other four are in the cytosol. Defects in any of these can lead to some form of porphyria.

The hepatic porphyrias are characterized by acute neurological attacks (seizures, psychosis, extreme back and abdominal pain and an acute polyneuropathy), while the erythropoietic forms present with skin problems, usually a light-sensitive blistering rash and increased hair growth.

Variegate porphyria (also *porphyria variegata* or *mixed porphyria*), which results from a partial deficiency in PROTO oxidase, manifests itself with skin lesions similar to those of porphyria cutanea tarda combined with acute neurologic attacks. All other porphyrias are either skin- or nerve-predominant.

Porphyrin Coordinated to Iron: Heme

A heme (American English) or haem (British English) is a prosthetic group that consists of an iron atom contained in the centre of a large heterocyclic organic ring called a *porphyrin*. Not all porphyrins contain iron, but a substantial fraction of porphyrin-containing metalloproteins have heme as their prosthetic group; these are known as hemoproteins. Hemes are most commonly recognized in their presence as components of hemoglobin but they are also components of a number of other hemoproteins.

Function

Hemoproteins have diverse biological functions including the transportation of diatomic gases, chemical catalysis, diatomic gas detection, and electron transfer. The heme iron serves as a source or sink of electrons during electron transfer or redox chemistry. In peroxidase reactions, the porphyrin molecule also serves as an electron source. In the transportation or detection of diatomic gases, the gas binds to the heme iron. During the detection of diatomic gases, the binding of the gas ligand to the heme iron induces conformational changes in the surrounding protein.

It has been speculated that the original evolutionary function of hemoproteins was electron transfer in primitive sulfur-based photosynthesis pathways in ancestral cyanobacteria before the appearance of molecular oxygen.

Hemoproteins achieve their remarkable functional diversity by modifying the environment of the heme macrocycle within the protein matrix. For example, the ability of hemoglobin to effectively deliver oxygen to tissues is due to specific amino acid residues located near the heme molecule. Hemoglobin binds oxygen in the pulmonary vasculature, where the pH is high and the pCO_2 is low, and releases it in the tissues, where the situations are reversed. This phenomenon is known as the Bohr effect. The molecular mechanism behind this effect is the steric organization of the globin chain; a histidine residue, located adjacent to the heme group, becomes positively charged under acid (low pH) circumstances (which are caused by dissolved CO_2 in working muscles, etc.), sterically releasing oxygen from the heme group.

Other Hemes

Heme L is the derivative of heme B which is covalently attached to the protein of lactoperoxidase, eosinophil peroxidase and thyroid peroxidase. The addition of peroxide with the glutamyl-375 and aspartyl-225 of lactoperoxidase forms ester bonds between these amino acid residues and the heme 1- and 5-methyl groups, respectively. Similar ester bonds with these two methyl groups are thought to form in eosinophil and thyroid peroxidases. Heme L is one important characteristic of animal peroxidases; plant peroxidases incorporate heme B. Lactoperoxidase and eosinophil peroxidase are protective enzymes responsible for the destruction of invading bacteria and virus. Thyroid peroxidase is the enzyme catalysing the biosynthesis of the important thyroid hormones. Because lactoperoxidase destroys invading organisms in the lungs and excrement, it is thought to be an important protective enzyme.

Heme M is the derivative of heme B covalently bound at the active site of myeloperoxidase. Heme M also contains the two ester bonds at the heme 1- and 5-methyls, much as the other mammalian peroxidases. In addition, a unique sulfonium ion linkage between the sulfur of a methionyl aminoacid residue and the heme 2-vinyl group is formed, giving this enzyme the unique capability of easily oxidizing chloride and bromide ions. Myeloperoxidase is present in mammalian neutrophils and is responsible for the destruction of invading bacteria and virus. It also synthesizes hypobromite by “mistake” which is a known mutagenic compound.

Heme D is another derivative of heme B, but in which the propionic acid side chain at the carbon of position 6, ring III is bound to this carbon both via the usual C-C bond but also by the carboxyl oxygen, giving heme D a fifth ring and a lactone. Ring III is also hydroxylated at position 5, in a conformation trans to the new lactone group. Heme D is the site for oxygen reduction to water of many types of bacteria at low oxygen tension.

Heme S is related to heme B by the having a formyl group at position 2 in place of the 2-vinyl group. Heme S is found in the hemoglobin of marine worms. The correct structures of heme B and heme S were first elucidated by German chemist Hans Fischer.

The names of cytochromes typically (but not always) reflect the kinds of hemes they contain: cytochrome a contains heme A, cytochrome c contains heme C, etc.

Synthesis

The enzymatic process that produces heme is properly called porphyrin synthesis, as all the intermediates are tetrapyrroles that are chemically classified as porphyrins. The process is highly conserved across biology. In humans, this pathway serves almost exclusively to form heme. In other species, it also produces similar substances such as cobalamin (vitamin B_{12}).

The pathway is initiated by the synthesis of D-Aminolevulinic acid (dALA or δALA) from the amino acid glycine and succinyl-CoA from the citric acid cycle (Krebs cycle). The rate-limiting enzyme responsible for this reaction, *ALA synthase*, is strictly regulated by intracellular iron levels and heme concentration. A low-iron level, e.g., in iron deficiency, leads to decreased porphyrin synthesis, which prevents accumulation of the toxic intermediates. This mechanism is of therapeutic importance: infusion of *heme arginate* or *hematin* can abort attacks of porphyria in patients with an inborn error of metabolism of this process, by reducing transcription of ALA synthase. The organs mainly involved in heme synthesis are the liver and the bone marrow, although every cell requires heme to function properly. Heme is seen as an intermediate molecule in catabolism of haemoglobin in the process of bilirubin metabolism.

Degradation

Degradation begins inside macrophages of the spleen, which remove old and damaged erythrocytes from the circulation. In the first step, heme is converted to biliverdin by the enzyme heme oxygenase (HOXG). NADPH is used as the reducing agent, molecular oxygen enters the reaction.

In addition, heme degradation appears to be an evolutionary conserved response to oxidative stress. Briefly, when cells are exposed to free radicals, there is a rapid induction of the expression of the stress responsive heme oxygenase-1 (Hmox1) isoenzyme that catabolizes heme. The reason why cells must increase exponentially their capability to degrade heme in response to oxidative stress remains unclear but this appears to be part of a cytoprotective response that avoids the deleterious effects of free heme.

Heme in Health and Disease

Under homeostasis, the reactivity of heme is controlled by its insertion into the "heme pockets" of hemoproteins. Under oxidative stress however, some hemoproteins, e.g. hemoglobin, can release their heme prosthetic groups. The non-protein-bound (free) heme produced in this manner becomes highly cytotoxic, most probably due to the Fe atom contained within its protoporphyrin IX ring, which can undergo Fenton chemistry to catalyse in an unfettered manner the production of free radicals. This property of free heme can sensitize a variety of cell types to undergo programmed cell death in response to pro-inflammatory agonists. This deleterious effect is thought to play an important role in the pathogenesis of certain inflammatory diseases such as malaria.

Nitrogen-substituted Porphyrins: Phthalocyanine

Phthalocyanine is an intensely coloured macrocyclic compound that is widely used in dyeing. Phthalocyanines form coordination complexes with most elements of the periodic table. These complexes are also intensely coloured and also are used as dyes.

Properties

Phthalocyanine, abbreviated H_2Pc, has several unusual properties. It and its complexes have low solubility in virtually all solvents. One litre of warm (40 °C) benzene dissolves less than a milligram of H_2Pc and CuPc. These compounds dissolve in sulfuric acid owing to the protonation of the nitrogen centres that link the pyrrole rings. They are also highly stable thermally and typically resist melting. CuPc sublimes at >500 °C under one atmosphere of nitrogen.

Relevant to their main application, phthalocyanines strongly absorb light in the red portion of the optical spectrum (about 600 to 700 nm), thus these dyes are characteristically blue or greenish. Phthalocyanines are structurally related to other macrocyclic pigments, especially the porphyrins.

Both feature four pyrrole-like subunits linked to form a 16-membered ring. The pyrrole-like rings within H_2Pc are closely related to isoindole. Both porphyrins and phthalocyanines function as planar tetradentate dianionic ligands that bind metals through four inwardly projecting nitrogen centres. Such complexes are formally derivatives of Pc^{2-}, the conjugate base of H_2Pc.

Many derivatives of the parent phthalocyanine are known. In addition to the ring-substituted derivatives, there also exist subphthalocyanines, superphthalocyanine, and hemiporphyrazine.

History

An unidentified blue compound, which we now know was metal-free phthalocyanine, was described in 1907. In 1927, Swiss researchers accidentally synthesized copper phthalocyanine, copper naphthalocyanine, and copper octamethylphthalocyanine in an attempted conversion of *o*-dibromobenzene into phthalonitrile. They remarked on the enormous stability of these complexes but did not further characterize these blue complexes. The same blue product was further investigated at Scottish Dyes, Ltd., Grangemouth, Scotland (later ICI). ColorantHistory.org hosts an old documentary about the discovery of the pigment.

Synthesis

Phthalocyanine forms upon heating phthalic acid derivatives that contain nitrogen functional groups. Classical precursors are phthalonitrile and diiminoisoindole. In the presence of urea, the heating of phthalanhydride is a useful route to H_2Pc. These reactions are more efficient in the presence of metal salts. Other precursors include, o-cyanobenzamide,, and phthalimide. Using such methods, approximately 57000 tons of various phthalocyanines were produced in 1985.

Applications

Approximately 25% of all artificial organic pigments are phthalocyanine derivatives. Copper phthalocyanine (CuPc) dyes are produced by introducing solubilizing groups, such as one or more sulfonic acid functions. These dyes find extensive use in various areas of textile dyeing (Direct dyes for cotton), for spin dyeing and in the paper industry. Direct blue 86 is the sodium salt of CPC-sulfonic acid whereas direct blue 199 is the quaternary ammonium salt of the CPC-sulfonic acid. The quaternary ammonium salts of these sulfonic acids used as solvent dyes because of their solubility in organic solvents, e.g. Solvent Blue 38 and Solvent Blue 48. The dye derived from cobalt phthalocyanine and an amine is Phthalogen Dye IBN. 1,3-Diiminoisoindolene, the intermediate formed during phthalocyanine manufacture, used in combination with a copper salt affords the dye GK 161. All major artists' pigment manufacturers produce variants of copper phthalocyanine, designated colour index PB15 (blue) and colour indexes PG7 and PG36 (green).

Phthalocyanine is also commonly used as a dye in the manufacture of high-speed CD-R media. Most brands of CD-R use this dye except Taiyo Yuden and Verbatim DataLife (which use cyanine and azo dyes, respectively).

Niche Applications

Metal phthalocyanines have long been examined as catalysts for redox reactions. Areas of interest are the oxygen reduction reaction and the sweetening of gas streams by removal of hydrogen sulfide.

Phthalocyanine compounds have been investigated as donor materials in molecular electronics, e.g. organic field-effect transistors.

Porphin

Porphin, sometimes spelled porphine, is the parent chemical compound for types of biochemically significant compounds called porphyrins. The chemical formula of porphin is $C_{20}H_{14}N_4$. Porphin is an organic compound that is aromatic and heterocyclic since its chemical structure.

Structural Characteristics of Porphin

Around the perimeter of the macrocycle ring, there is a cyclic chain of sp hybridized carbon atoms, all of which are part of a conjugated double bond system, giving the molecule its aromatic character.

The aromatic character of porphin stems both from its conjugation as well as its flat or planar geometry, meaning that all the atoms lie in a single plane. Bonded to the cyclic chain of carbons are four nitrogen atoms facing the centre of the molecule, two of which are bonded to hydrogen atoms and the other two nitrogens forming part of the conjugated double bond system.

Complex with Metal in the Centre

If the two nitrogens give up their hydrogens, the four central nitrogen atoms can act as ligands to bond to a metal ion in the centre of the molecule. This is often the case with biological porphyrin compounds, especially with an iron atom in the centre. When there is no metal ion (or atom) bound to the nitrogens in the centre, then the compounds are called *free porphin* or *free porphyrins*. If they are bonded to a metal in the centre, then they are *bound*. A porphyrin with an iron atom of the type found in myoglobin, hemoglobin, or certain cytochromes is called heme.

Compounds similar to porphin are the parent compounds for similar ring systems with other central metal atoms in biochemistry. These include chlorin, which has a magnesium atom in several types of chlorophyll; bacteriochlorin, found in bacteriochlorophyll; and corrin, which has a cobalt atom in cobalamin or vitamin B12.

Bibliography

Alexandrov, S.: *Methods for Dissolution, Separation and Preconcentration in Analytical Chemistry,* Narodna Kultura, Sofia, 1995.

Alexandrov, S.; Christova, R.; D. Tsalev, B. Jeliazkova, V. Michaylova: *Practice in Quantitative Analysis*, University Press, Ohridski", 2000.

Arnaudov, M.: *Application of Infrared Spectroscopy in Chemistry,* Nauka i Izkustvo, Sofia, 1981..

Balkwill, Dr. Fran: *DNA Is Here to Stay*, Carolrhoda Books, Inc., Minneapolis, 1993.

Banwell, Colin and Mc. Cash Elaine: *Fundamentals of Molecular Spectroscopy*, Mc Graw-Hill, NY, 1994.

Bard, A.J.; Faulkner, L.R.: *Electrochemical Methods: Fundamentals and Applications*, John Wiley & Sons, New York, 2000.

Bast, R.C. Jr.: *Cancer Medicine*, Decker, Hamilton, 2000.

Bond, A. Curtis: *Modern Polarographic Methods in Analytical Chemistry*, M. Dekker, New York, 1991.

Bontchev, P.; I. Kuleff, M. Mitewa: *Department of Analytical Chemistry 1924–1994 - Bibliography*, University Press, Sofia, 1994.

Bontchev, P.R., *Complexation and Catalytic Activity*, Nauka i Izkustvo, Sofia, 1972.

Bummer, P. M.: *The Science and Practice of Pharmacy*, Mack, Easton, PA, 2000.

Burns, G.; Glazer, A.M.: *Space Groups for Scientists and Engineers,* Academic Press, Boston, 1990.

Callen, H. B.: *Thermodynamics*, Wiley, New York, 1960.

Christian, G. D.: *Analytical Chemistry*, Wiley, NY, 1992.

Christova, R., S. Alexandrov, D. Tsalev, B. Jeljazkova, V. Mihajlova: *Practice in Quantitative Analysis*, Ohridski University Press, Sofia, 2003.

Christova, R.; V. Simeonov: *Potentiometry*, Nauka i Izkustvo, Sofia, 1983.

Crouch, Stanley; Skoog, Douglas A.: *Principles of Instrumental Analysis*, Thomson Brooks/Cole, Australia, 2007.

Cussler, E. L.: *Diffusion: Mass Transfer in Fluid Systems*, Cambridge University Press, Cambridge, UK, 2009.

Dahmen, E. A. M. F.: *Electroanalysis: Theory and Applications in Aqueous and Non-aqueous Media and in Automated Chemical Control*, Elsevier, Amsterdam, 1986.

Dedina, J.; D.L. Tsalev: *Hydride Generation Atomic Absorption Spectrometry*, John Wiley & Sons, Chichester, 1995.

Denbigh, K.: *The Principles of Chemical Equilibrium*, Cambridge Univ. Press, London, 1966.

Dimov, N. K.; Lekova, E. Russeva, I. Havezov, D. Tsalev: *English-Bulgarian Glossary. Instrumental Methods of Analysis*, NIHFI (Pharmaceutical Research Institute), Sofia, 1996.

Duarte, F. J. and L.W. Hillman: *Dye Laser Principles,* Academic, New York, 1990.

Duxbury, G.: *Infrared Vibration-Rotation Spectroscopy, From Free Radicals to the Infrared Sky*, Wiley, NJ, 1999.

Evans, A.: *Potentiometry and Ion Selective Electrodes*, Wiley, New York, 1987.

Gardiner, W. P.: *Statistical Analysis Methods for Chemists: A software-based approach*, Royal Society of Chemistry, Cambridge, UK, 1997.

Guyton, A.C.; Hall, J.E.: *Textbook of Medical Physiology*, Saunders, Philadelphia, 2006.

Hala, E., I. Wichterle, J. Polak, and T. Boublik: *Vapor–Liquid Equilibrium at Normal Pressures,* Pergamon, Oxford, 1968.

Havezov, I.P.; D.L. Tsalev: *Atomic Absorption Analysis*, Nauka i Izkustvo, Sofia, 1980.

Holland, C. D.: *Fundamentals of Multicomponent Distillation,* McGraw-Hill, New York, 1981.

Holler, F. James; Skoog, Douglas A., West, Donald M.: *Fundamentals of Analytical Chemistry*, Saunders College Pub., Philadelphia, 1996.

Horsely, L. H.: *Azeotropic Data, ACS Advances in Chemistry*, American Chemical Society, Washington, DC, 1952.

Hubert H. Girault: *Analytical and Physical Electrochemistry*, EPFL. , Lausanne, 2004.

Humphrey, J. L,. and G. E. Keller II: *Separation Process Technology*, McGraw-Hill, New York, 1997.

Jeliazkova, B.; Ch. Daiev, S. Beltchev, D. Tsalev, L. Shishkova, M. Mitewa, Yu. Charizanov, N. Piperov: *Practice in Physical Methods of Analysis and Investigation of Inorganic Materials*, University Press, Ohridski, 1991.

Jones, Denny: *Principles and Prevention of Corrosion,* Upper Saddle River, Prentice Hall, New Jersey, 1996.

Jornitz, Maik W.: *Sterile Filtration*, Springer, Germany, 2006.

Joseph Needham: *Science & Civilisation in China,* Cambridge University Press, Cambridge, England, 1962.

Joseph, C. Salamone: *Polymeric Materials Encyclopedia,* CRC Press, Boca Raton, FL, 1996.

Karger, B. L., L. R. Snyder, and C. Horvath: *An Introduction to Separation Science*, John Wiley and Sons, New York, 1973.

Kenneth I. Ozomwna: *Recent Advances in Analytical Electrochemistry 2007*, Transworld Research Network, 2007.

Kittel and Kroemer: *Thermal Physics*, W. H. Freeman, New York, 1980.

Kotcheva, L.L.: D. Chr. Nonova: *Qualitative Semi-Micro Analysis*, University Press, Ohridski, Sofia, 2000.

Kuleff, I.; R. Djingova: *75th Anniversary of the Department of Analytical Chemistry 1924–1994 - Bibliography 1994–1999* , University Press, Sofia, 1999.

Kuleff, I.; R. Zlatanova, P. Konstantinov, M. Dimitrov, G.Genchev , P. Peev: *How Dangerous for the Population Are the Liquid Radioactive Releases from NPP "Kozloduy"?*, BalBok, Sofia, 1994..

Laidler, Keith: *The World of Physical Chemistry*, Oxford University Press, Oxford, 1993.

Lindon, John C.: *Encyclopedia of Spectroscopy and Specrometry*, Academic Press, Uk, 1999.

Massart, D. L.: *Chemometrics: A Textbook*, Amsterdam, Elsevier, 1988.

Miller, J. N. and J. C. Miller: *Statistics and Chemometrics for Analytical Chemistry*, Prentice Hall, UK, 2000.

Nicholas P.: *Hazardous Chemicals in the Polymer Industry,* M. Dekker, Cheremisinoff, New York, 1995.

Nieman, Timothy A.; Skoog, Douglas A.; Holler, F. James: *Principles of Instrumental Analysis*, Brooks/Cole, Pacific Grove, CA, 1998.

Nikolov, G.; P.R. Bontchev: *Mechanisms of Inorganic Reactions in Solutions*, Nauka i Izkustvo, Sofia, 1969.

Norman, A. Waterman and Michael F. Ashby.: *The Materials Selector,* CRC Press, Boca Raton, 1997.

Odian, George: *Principles of Polymerization,* John Wiley & Sons, New York, 1991.

Olagoke Olabisi: *Handbook of Thermoplastics,* Marcel Dekker, New York, 1997.

Perry, R. H. and D. Green: *Perry's Chemical Engineer's Handbook*, McGraw-Hill, New York, 1984.

Prausnitz, J. M., R. N. Lichtenthaler, and E. G. de Azevedo: *Molecular Thermodynamics of Fluid-Phase Equilibria*, Prentice-Hall, Englewood Cliffs, NJ, 1986.

Prichard, E.: *Analytical Measurement Terminology*, LGC/Royal Society of Chemistry, Cambridge, UK, 2001.

Reid, R. C., J. M. Prausnitz, and B. E. Poling: *The Properties of Gases and Liquids*, McGraw-Hill, New York, 1987.

Seeman, N.C.: *Biomolecular Stereodynamics*, Adenine Press, New York, 1981.

Simeonov, V.: *Information Aspects in Chemical Analysis*, Obrazovatelni Tekhnologii, Sofia, 1999.

Skoog, D.A.: *Principles of Instrumental Analysis*, Saunders College Publishing, 1991.

Spasov, St.; M. Arnaudov: *Application of Spectroscopy in Organic Chemistry*, Nauka i Izkustvo, Sofia, 1978.

Stone, D. C.: *Instrumental Analysis Laboratory Manual*, Department of Chemistry, University of Toronto, 2005.

Stuart, M. Lee: *International Encyclopedia of Composites,* VCH Publishers, New York, 1990.

Tsalev, D.L.: *Atomic Absorption Spectrometry in Occupational and Environmental Health Practice*, Vol. II: *Determination of Individual Elements*, CRC Press, Boca Raton, Florida, 1984.

Tsalev, D.L.; I.P. Havezov: *Atomic Absorption Analysis*, Khimia, Leningrad, 1983.

Walas, S. M.: *Phase Equilibria in Chemical Engineering,* Butterworths, Reading, MA, 1985.

Wang, Joseph C.: *Analytical Electrochemistry*, John Wiley & Sons, Chichester, 2000.

Workman, Jerome; Workman, Jerry: *Handbook of Organic Compounds: NIR, IR, Raman, and UV-Vis Spectra Featuring Polymers and Surfactants*, Academic Press, San Diego, 2001.

Zaprianov, Z.K.: *Atomic Absorption Spectrometry in Occupational and Environmental Health Practice*, Vol. I: *Analytical Aspects and Health Significance*, CRC Press, Boca Raton, Florida, 1983.

Index

T

U

V

X

□□□